〔大数据技术丛书〕

Hadoop
大数据分析技术

迟殿委 陈鹏程 主编

清华大学出版社
北京

内 容 简 介

伴随 Hadoop 的成长，Hadoop 不再是一个简单的数据分布式存储平台和工具，已经成长为一个完整的生态圈。本书采用 Hadoop 3.2.2 版本，系统讲解 Hadoop 生态系统主流的大数据分析技术。本书配套示例源码、PPT 课件、教学大纲与编程环境。

本书共分 11 章。内容包括 Hadoop 概述与大数据环境准备、Hadoop 伪分布式集群搭建、HDFS 分布式存储实战、MapReduce 实战、ZooKeeper 与高可用集群实战、Hive 数据仓库实战、HBase 数据库实战、Flume 数据采集实战、Kafka 实战、影评大数据分析项目实战、旅游酒店评价大数据分析项目实战。

本书可作为 Hadoop 大数据技术初学者的入门书，也可作为 Hadoop 大数据分析工程师的指导手册，还可作为高等院校或者高职高专大数据专业的教材或教学参考书。

本书封面贴有清华大学出版社防伪标签，无标签者不得销售。
版权所有，侵权必究。举报：010-62782989，beiqinquan@tup.tsinghua.edu.cn。

图书在版编目（CIP）数据

Hadoop 大数据分析技术 / 迟殿委，陈鹏程主编. —北京：清华大学出版社，2022.11
（大数据技术丛书）
ISBN 978-7-302-62099-0

Ⅰ. ①H… Ⅱ. ①迟… ②陈… Ⅲ. ①数据处理软件 Ⅳ. ①TP274

中国版本图书馆 CIP 数据核字（2022）第 198810 号

责任编辑：夏毓彦
封面设计：王　翔
责任校对：闫秀华
责任印制：杨　艳

出版发行：清华大学出版社
网　　址：http://www.tup.com.cn，http://www.wqbook.com
地　　址：北京清华大学学研大厦 A 座　　**邮　编**：100084
社 总 机：010-83470000　　**邮　购**：010-62786544
投稿与读者服务：010-62776969，c-service@tup.tsinghua.edu.cn
质量反馈：010-62772015，zhiliang@tup.tsinghua.edu.cn

印 装 者：三河市铭诚印务有限公司
经　　销：全国新华书店
开　　本：190mm×260mm　　**印　张**：15.5　　**字　数**：418 千字
版　　次：2022 年 11 月第 1 版　　**印　次**：2022 年 11 月第 1 次印刷
定　　价：69.00 元

产品编号：088513-01

前　言

国家提出要加快 5G 网络和数据中心等新型基础设施建设（简称新基建）的进度。其中，信息化新型基础设施包含云计算、大数据、人工智能、区块链、5G 等内容。大数据是指具有海量（volume）、多模态（variety）、变化速度快（velocity）、蕴含价值高（value）和真实性（veracity）"5V"特征的数据，使得传统的数据存储、管理、分析技术已经无法满足大数据的处理要求。大数据给传统的数据处理和数据分析带来巨大的挑战，已引起学术界和工业界的高度关注。Hadoop 正是在这种背景下产生的一个大数据开源平台。许多大型互联网公司，如谷歌、阿里巴巴、百度、京东等互联网公司都急需掌握 Hadoop 大数据技术的人才，而目前人才市场上大数据技术相关人才由于种种原因存在供不应求的状况，本书在这个背景下创作而成。

本书内容

本书是一本关于 Hadoop 3.2.2 大数据平台搭建和数据分析、生态体系主要组件的应用和开发方面的实战书籍，涉及的知识面比较广，涵盖了当前整个 Hadoop 生态系统主流的大数据开发技术。本书从实践操作与开发讲起，在基本操作已经掌握以后，再回过头来讲解理论知识。所以，本书是先实践再理论，方便读者快速掌握 Hadoop 大数据分析技术。

全书共分 11 章，第 1 章讲解 Hadoop 框架简介及新版本特性，并详细介绍大数据环境的准备工作，包括 Linux 操作系统的安装、SSH 工具使用和配置等；第 2 章讲解 Hadoop 伪分布式的安装和开发体验，使读者熟悉 Hadoop 大数据开发两大核心组件，即 HDFS 和 MapReduce；第 3~9 章讲解 Hadoop 生态系统各框架 HDFS、MapReduce、输入/输出、Hadoop 集群配置、ZooKeeper、HBase、Hive、Flume 数据采集系统、Kafka 等，并通过实际案例加深对各个框架的理解与应用。第 10~11 章分别通过影评大数据分析项目实战和旅游酒店评价大数据分析项目实战，使读者了解完整的大数据项目开发过程，并巩固所学的知识，使之掌握的内容更加系统、全面。

本书目的

通过本书的学习，读者可以对照书中的步骤成功搭建属于自己的 Hadoop 大数据集群，并掌握基于 Hadoop 的大数据分析与开发技术，最终能够独立完成 Hadoop 大数据分析与开发项目。

本书适合的读者

本书可作为 Hadoop 框架初学者的入门书以及大数据分析人员的参考手册，也可作为高校开设大数据平台搭建或大数据开发课程的参考教材。学习本书要求读者有一定的 Java 编程基础并了解 Linux 系统的基础知识。本书每一个章节的实践操作内容都有详细清晰的步骤讲解，即使读者没有任何大数据基础，也可以对照书中的步骤成功搭建属于自己的大数据集群，本书是一本真正提高读者动手能力、以实操为主的入门书籍。通过本书的学习，结合每章配套的源代码，读者能够迅速理解与掌握 Hadoop 大数据相关技术框架，并可以熟练使用 Hadoop 集成环境进行大数据项目的开发。

配套源码、PPT 课件等资源下载

本书配套源码、PPT 课件、教学大纲与编程环境，需要用微信扫描下边二维码获取，可按扫描后的页面提示填写你的邮箱，把下载链接转发到邮箱中下载。如果下载有问题或阅读中发现问题，请联系 booksaga@163.com，邮件主题写"Hadoop 大数据分析技术"。

作　者
2022 年 9 月

目 录

第1章 Hadoop 概述与大数据环境准备 1
    1.1 大数据定义 2
    1.2 Hadoop 生态介绍 2
        1.2.1 Hadoop 简介 2
        1.2.2 Hadoop 版本简介 4
        1.2.3 Hadoop 生态系统和组件介绍 6
    1.3 Hadoop 3 新特性 7
    1.4 虚拟机安装 9
    1.5 安装 Linux 操作系统 10
    1.6 SSH 工具与使用 15
    1.7 Linux 统一设置 16
    1.8 小结 18

第2章 Hadoop 伪分布式集群搭建 19
    2.1 安装独立运行的 Hadoop 19
    2.2 Hadoop 伪分布式环境准备 22
    2.3 Hadoop 伪分布式安装 26
    2.4 HDFS 操作命令 31
    2.5 Java 项目访问 HDFS 33
        2.5.1 创建 Maven 项目 34
        2.5.2 HDFS 操作示例 36
    2.6 winutils 38
    2.7 快速 MapReduce 程序示例 39
    2.8 小结 42

第3章 HDFS 分布式存储实战 43
    3.1 HDFS 的体系结构 43
    3.2 NameNode 的工作 44
        3.2.1 查看镜像文件 45
        3.2.2 查看日志文件 46
        3.2.3 日志文件和镜像文件的操作过程 47
    3.3 SecondaryNameNode 49
    3.4 DataNode 50

- 3.5 HDFS 的命令 ... 50
- 3.6 远程过程调用 ... 51
- 3.7 小结 ... 53

第 4 章 MapReduce 实战 ... 55

- 4.1 MapReduce 的运算过程 ... 55
- 4.2 WordCount 示例 ... 57
- 4.3 自定义 Writable ... 60
- 4.4 Partitioner 分区编程 ... 64
- 4.5 自定义排序 ... 66
- 4.6 Combiner 编程 ... 67
- 4.7 默认 Mapper 和默认 Reducer ... 68
- 4.8 倒排索引 ... 69
- 4.9 Shuffle ... 73
- 4.10 小结 ... 77

第 5 章 ZooKeeper 与高可用集群实战 ... 79

- 5.1 ZooKeeper 简介 ... 79
  - 5.1.1 Zxid ... 80
  - 5.1.2 版本号 ... 81
- 5.2 单一节点安装 ZooKeeper ... 82
- 5.3 基本客户端命令 ... 83
- 5.4 Java 代码操作 ZooKeeper ... 86
- 5.5 ZooKeeper 集群安装 ... 91
- 5.6 znode 节点类型 ... 92
- 5.7 观察节点 ... 93
- 5.8 配置 Hadoop 高可用集群 ... 93
- 5.9 用 Java 代码操作集群 ... 102
- 5.10 小结 ... 104

第 6 章 Hive 数据仓库实战 ... 105

- 6.1 Hive3 的安装配置 ... 107
- 6.2 Hive 的命令 ... 110
- 6.3 Hive 内部表 ... 114
- 6.4 Hive 外部表 ... 116
- 6.5 Hive 表分区 ... 117
  - 6.5.1 分区技术细节 ... 117
  - 6.5.2 分区示例 ... 119
- 6.6 查询示例汇总 ... 121
- 6.7 Hive 函数 ... 122
- 6.8 Hive 自定义函数 ... 128
- 6.9 Hive 视图 ... 132

		6.10	hiveserver2 132

	6.11	使用 JDBC 连接 hiveserver2 134
	6.12	小结 135

第 7 章 HBase 数据库实战 136

	7.1	HBase 的特点 136
	7.2	HBase 安装 139
		7.2.1 HBase 的单节点安装 140
		7.2.2 HBase 的伪分布式安装 142
		7.2.3 Java 客户端代码 144
	7.3	HBase 集群安装 150
	7.4	HBase Shell 操作 153
		7.4.1 数据模型定义 154
		7.4.2 数据基本操作 156
	7.5	协处理器 160
	7.6	Phoenix 162
	7.7	小结 168

第 8 章 Flume 数据采集实战 169

	8.1	Flume 的安装与配置 170
	8.2	快速示例 171
	8.3	在 ZooKeeper 中保存 Flume 的配置文件 172
	8.4	Flume 的更多 Source 176
		8.4.1 Avro Source 176
		8.4.2 Thrift Source 和 Thrift Sink 180
		8.4.3 Exec Source 183
		8.4.4 Spool Source 184
		8.4.5 HDFS Sinks 184
	8.5	小结 185

第 9 章 Kafka 实战 186

	9.1	Kafka 的特点 187
	9.2	Kafka 术语 188
	9.3	Kafka 安装与部署 189
		9.3.1 单机部署 189
		9.3.2 集群部署 195
	9.4	小结 198

第 10 章 影评大数据分析项目实战 199

	10.1	项目介绍 199
	10.2	项目需求分析 199
	10.3	项目详细实现 203

- 10.3.1 搭建项目环境 ... 203
- 10.3.2 编写爬虫类 ... 206
- 10.3.3 编写分词类 ... 207
- 10.3.4 第一个 job 的 Map 阶段实现 ... 210
- 10.3.5 第一个 job 的 Reducer 阶段实现 ... 210
- 10.3.6 第二个 job 的 Map 阶段实现 ... 211
- 10.3.7 第二个 job 的自定义排序类阶段的实现 ... 211
- 10.3.8 第二个 job 的自定义分区阶段实现 ... 212
- 10.3.9 第二个 job 的 Reduce 阶段实现 ... 212
- 10.3.10 Run 程序主类实现 ... 213
- 10.3.11 编写词云类 ... 214
- 10.3.12 效果测试 ... 215

第 11 章 旅游酒店评价大数据分析项目实战 ... 216

11.1 项目介绍 ... 216
11.2 项目需求分析 ... 217
- 11.2.1 数据集需求 ... 217
- 11.2.2 功能需求 ... 217

11.3 项目详细实现 ... 218
- 11.3.1 数据集上传到 HDFS ... 219
- 11.3.2 Hadoop 数据清洗 ... 221
- 11.3.3 构建 Hive 数据仓库表 ... 225
- 11.3.4 Sqoop 数据导入与导出 ... 230
- 11.3.5 数据可视化开发 ... 232

第 1 章

Hadoop概述与大数据环境准备

主要内容：

- ❖ 大数据概念。
- ❖ Hadoop简介。
- ❖ 虚拟机的安装与配置。
- ❖ Linux的操作系统的安装。
- ❖ SSH（Secure Shell）。

本章首先介绍大数据的基础知识，然后对Hadoop框架进行详细介绍，最后讲解Hadoop平台集群搭建的准备工作。Hadoop是由Apache基金会开发的分布式系统基础架构，也是Apache软件基金会的顶级开源项目，它的logo如图1-1所示。Hadoop的作者为Doug Cutting，他也是Lucene、Nutch等项目的创始人。2004年，Cutting基于Google（谷歌）发布的关于GFS（Google File System）学术文献打造出了Hadoop。

图1-1 Hadoop logo

Hadoop的特点在于用户可以在不了解分布式底层细节的情况下，开发分布式程序，充分利用集群的威力进行高速运算和存储。

Hadoop实现了一个分布式文件系统（Hadoop Distributed File System，简称HDFS）。HDFS有高容错性的特点，用来部署在低廉（low-cost）的硬件上，而且它提供高吞吐（high throughput）量来访问应用程序的数据，适合那些有着超大数据集（large data set）的应用程序。HDFS放宽了可移植操作系统接口的要求，可以以流的形式访问文件系统中的数据。

Hadoop框架最核心的设计就是HDFS和MapReduce。HDFS为海量的数据提供了存储，而MapReduce为海量的数据提供了计算。

1.1 大数据定义

大数据（Big data）又称为巨量资料、巨量数据或海量数据。一般来说，大数据的特性可概括为4V，即Volume、Variety、Velocity、Value。

1. Volume（大量数据）

- 累积庞大的数据：因特网、企业IT、物联网、社区、短信、电话、网络搜索、在线交易等，随时都在快速累积庞大的数据。
- 数据量等级：数据量很容易达到TB（Terabyte，1024GB），甚至PB（Petabyte，1024TB）或EB（Exabyte，1024PB）的等级。

2. Variety（多样性）

大数据的数据类型非常多样化，可分为非结构化信息和结构化信息。

- 非结构化信息：文字、图片、图像、视频、音乐、地理位置信息、个人化信息——如社区、交友数据等。
- 结构化信息：数据库、数据仓库等。

3. Velocity（时效性）

- 数据的传输流动：随着带宽越来越大、设备越来越多，每秒产生的数据流越来越大。
- 必须能实时处理大量的信息：时间太久就会失去数据的价值，所以数据必须能在最短时间内分析出结果。

4. Value（价值密度低）

大数据价值密度相对较低。如随着物联网的广泛应用，信息感知无处不在，信息海量，但价值密度较低，存在大量不相关信息。因此需要对未来趋势与模式作预测分析，利用机器学习、人工智能等进行深度复杂分析。而如何通过强大的机器算法更迅速地完成数据的价值提炼，是大数据时代急需解决的难题。虽然单位数据的价值密度在不断降低，但是数据的整体价值在提高。

大数据的影响已经深入到各个领域和行业，在商业、经济及其他领域中，将大量数据进行分析后就可得出许多数据的关联性，可用于预测商业趋势、营销研究、金融财务、疾病研究、打击犯罪等。决策行为将基于数据和分析的结果，而不是依靠经验和直觉。

1.2 Hadoop 生态介绍

1.2.1 Hadoop 简介

Hadoop起源于Apache Nutch，后者是一个开源的网络搜索引擎，本身也是Lucene项目的一部分。

Hadoop这个名字不是一个缩写,它是一个虚构的名字。该项目的创建者Doug Cutting如此解释Hadoop的得名:"这个名字是我孩子给一头吃饱了的棕黄色大象起的名字。我的命名标准就是简短、容易发音和拼写,没有太多的意义,并且不会被用于别处。小孩子是这方面的高手。Googol就是由小孩命名的。"(Google来源于Googol一词。GooGol指的是10的100次幂(方),代表互联网上的海量资源。公司创建之初,肖恩·安德森在搜索该名字是否已经被注册时,将Googol误打成了Google。)

Hadoop及其子项目和后继模块所使用的名字往往也与其功能不相关,经常用一头大象或其他动物主题(例如Pig)。较小的各个组成部分给予更多描述性(因此也更通俗)的名称。这是一个很好的原则,因为它意味着可以大致从其名字猜测其功能,例如,jobtracker 的任务就是跟踪MapReduce作业。

从头开始构建一个网络搜索引擎是一个雄心勃勃的目标,不只是要编写一个复杂的、能够抓取和索引网站的软件,还需要面临着没有专业运行团队支持运行它的挑战,因为它有那么多独立部件。同样昂贵的还有:据Mike Cafarella和Doug Cutting估计,一个支持此10亿页的索引,需要价值约50万美元的硬件投入,每月运行费用还需要3万美元。不过,他们相信这是一个有价值的目标,因为这会开源并最终使搜索引擎算法普及化。

Nutch项目开始于2002年,一个可工作的抓取工具和搜索系统很快浮出水面。但他们意识到,他们的架构将无法扩展到拥有数十亿网页的网络。2003年发表的一篇描述Google分布式文件系统(简称GFS)的论文为他们提供了及时的帮助,文中称Google正在使用此文件系统。GFS或类似的东西,可以解决他们在网络抓取和索引过程中产生的大量文件的存储需求。具体而言,GFS会省掉管理所花的时间,如管理存储节点。在2004年,他们开始写一个开放源码的应用,即Nutch的分布式文件系统(NDFS)。

2004年,Google发表了论文,向全世界介绍了MapReduce。2005年年初,Nutch的开发者在Nutch上有了一个可工作的MapReduce应用,到了年中,所有主要的Nutch算法被移植到使用MapReduce和NDFS来运行。

Nutch中的NDFS和MapReduce实现的应用远不只是搜索领域,2006年2月,他们从Nutch转移出来成为一个独立的Lucene子项目,称为Hadoop。大约同一时间,Doug Cutting加入雅虎,Yahoo提供一个专门的团队和资源将Hadoop发展成一个可在网络上运行的系统(见后文的补充材料)。2008年2月,雅虎宣布其搜索引擎产品部署在一个拥有1万个内核的Hadoop集群上。

2008年1月,Hadoop已成为Apache顶级项目,证明它是成功的,它成为一个多样化、活跃的社区。通过这次机会,Hadoop成功地应用在雅虎之外的很多公司,如Last.fm、Facebook和《纽约时报》。一些应用在Hadoop维基网站上有介绍,Hadoop维基的网址为http://wiki.apache.org/hadoop/PoweredBy。

有一个良好的宣传范例,《纽约时报》使用亚马逊的EC2云计算将4TB的报纸扫描文档压缩,转换为用于Web的PDF文件。这个过程历时不到24小时,使用100台机器同时运行。如果不结合亚马逊的按小时付费的模式(即允许《纽约时报》在很短的一段时间内访问大量机器)和Hadoop易于使用的并行程序设计模型,该项目很可能不会这么快开始启动。

2008年4月,Hadoop打破世界纪录,成为最快排序1TB数据的系统,运行在一个910个节点的集群上,Hadoop在209秒内排序了1 TB的数据(还不到3.5分钟),击败了前一年费时297秒的冠军。同年11月,谷歌在报告中声称,它的MapReduce实现了执行1 TB数据的排序只用68秒。2009年5月,有报道称Yahoo的团队使用Hadoop对1 TB的数据进行排序,只花了62秒。

构建互联网规模的搜索引擎需要大量的数据,因此需要大量的机器来进行处理。Yahoo!Search包括四个主要组成部分:Crawler,从因特网下载网页;WebMap,构建一个网络地图;Indexer,为最佳页面构建一个反向索引；Runtime（运行时），回答用户的查询。WebMap是一幅图，大约包括一万亿条边（每条代表一个网络链接）和一千亿个节点（每个节点代表不同的网址）。创建和分析此类大图需要大量计算机运行若干天。2005年年初，WebMap所用的基础设施名为Dreadnaught，需要重新设计以适应更多节点的需求。Dreadnaught成功地从20个节点扩展到600个，但还需要一个完全重新的设计，以进一步扩展节点。Dreadnaught与MapReduce有许多相似的地方，但灵活性更强，结构更少。具体说来，Dreadnaught作业可以将输出发送到此作业下一阶段中的每一个分段（fragment），但排序是在库函数中完成的。在实际情形中，大多数WebMap阶段都是成对存在的，对应于MapReduce。因此，WebMap应用并不需要为了适应MapReduce而进行大量重构。

Eric Baldeschwieler（Eric14）组建了一个小团队，他们开始设计并原型化一个新的框架（原型为GFS和MapReduce，用C++语言编写），打算用它来替换Dreadnaught。尽管当务之急是需要一个WebMap新框架，但很显然，标准化对于整个Yahoo! Search平台至关重要，并且通过使这个框架泛化，足以支持其他用户，这样他们才能够充分运用其对整个平台的投资。

与此同时，雅虎在关注Hadoop（当时还是Nutch的一部分）及其进展情况。2006年1月，雅虎聘请了Doug Cutting，一个月后，决定放弃自己的原型，转而使用Hadoop。相较于雅虎自己的原型和设计，Hadoop的优势在于它已经在20个节点上实际应用过。这样一来，雅虎便能在两个月内搭建一个研究集群，并着手帮助真正需要的客户使用这个新的框架，速度比原来预计的要快许多。另一个明显的优点是Hadoop已经开源，较容易（虽然远没有那么容易）从雅虎法务部门获得许可。因此，雅虎在2006年初设立了一个200个节点的研究集群，他们将WebMap的计划暂时搁置，转而为研究用户支持和发展Hadoop。

1.2.2　Hadoop版本简介

目前市面上Hadoop版本主要有两种：Apache版本和第三方发行版本。Apache Hadoop是一款支持数据密集型分布式应用，并以Apache 2.0许可协议发布的开源软件框架。它支持在商品硬件构建的大型集群上运行的应用程序。Hadoop是根据Google公司发表的有关MapReduce和Google文件系统的论文设计的，称为社区版Hadoop。

第三方发行版Hadoop遵从Apache开源协议，用户可以免费地任意使用和修改Hadoop，也正是因此，市面上出现了很多Hadoop版本。其中有很多厂家在Apache Hadoop的基础上开发自己的Hadoop产品，比如Cloudera的CDH、Hortonworks的HDP、MapR的MapR产品等。

这两种版本各自优缺点如下：

1. Aapche版本的Hadoop

官方网址：http://hadoop.apache.org/。
Aapche Hadoop优势：对硬件要求低，完全开源免费，社区活跃，文档、资料翔实。
Aapche Hadoop劣势：搭建烦琐，维护烦琐，升级烦琐，添加组件烦琐。

2. 第三方发行版本的Hadoop

官方网址：https://www.cloudera.com/。

优势：

- 版本管理清晰。比如Cloudera CDH1、CDH2、CDH3、CDH4、CDH5等。后面加上补丁版本，如CDH4.1.0 patch level 923.142，表示在原生态Apache Hadoop 0.20.2基础上添加了1065个patch。
- 比Apache Hadoop在兼容性、安全性、稳定性上有增强。第三方发行版通常都经过了大量的测试验证，有众多部署实例，大量地运行在各种生产环境。
- 版本更新快。通常情况，比如CDH每个季度会有一个update，每一年会有一个release。
- 基于稳定版本Apache Hadoop，并应用了最新BUG修复或Feature的patch。
- 提供了部署、安装、配置工具，大大提高了集群部署的效率，可以在几个小时内部署好集群。
- 运维简单。提供了管理、监控、诊断、配置修改的工具，管理配置方便，定位问题快速、准确，运维工作简单、高效。

缺点：对硬件要求高。

3. CDH及其架构

第三方发行版本使用比较多的是CDH，其拥有最多的部署案例，提供强大的部署、管理和监控工具。Cloudera开发并贡献了可实时处理大数据的Impala项目，有强大的社区支持，当出现一个问题时，能够通过社区、论坛等网络资源快速获取解决方法。Cloudera Manager是一个管理CDH的端到端的应用，其主要作用包括：管理、监控、诊断、集成。

CDH的Hadoop版本集群中CDH管理界面如图1-2所示。

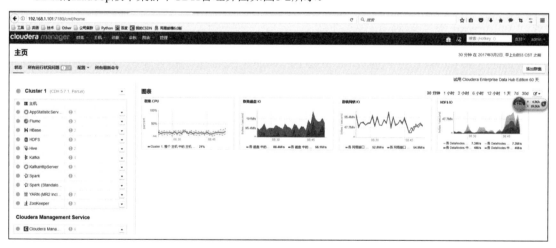

图1-2　CDH管理界面

CDH 架构如图1-3所示。

（1）Server

管理控制台服务器和应用程序逻辑。

负责软件安装、配置，启动和停止服务。

管理服务运行的集群。

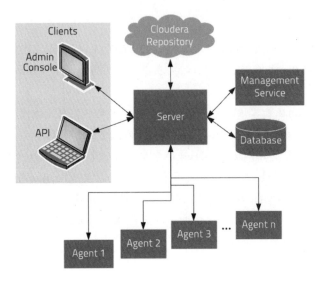

图1-3　CDH架构

（2）Agent

安装在每台主机上。

负责启动和停止进程，配置、监控主机。

（3）Management Service

由一组角色组成的服务，执行各种监视、报警和报告功能。

1.2.3　Hadoop生态系统和组件介绍

Hadoop生态系统组件主要包括：MapReduce、HDFS、HBase、Hive、Pig、ZooKeeper、Mahout、Flume、Sqoop，具体如图1-4所示。

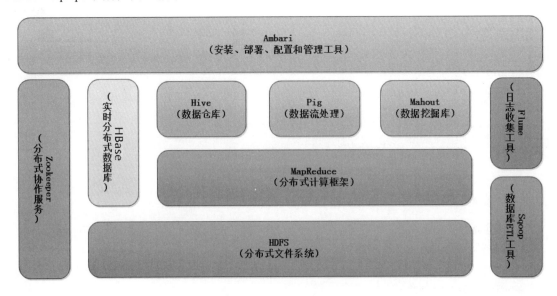

图1-4　Hadoop生态系统组件

下面具体介绍其中的主要组件：

- MapReduce：MapReduce是使用集群的并行、分布式算法处理大数据集的可编程模型。Apache MapReduce是从Google MapReduce派生而来的，用于在大型集群中简化数据处理。当前的Apache MapReduce版本基于Apache YARN框架构建的。YARN是"Yet-Another-Resource-Negotiator"的缩写。YARN可以运行非MapReduce模型的应用。YARN是Apache Hadoop想要超越MapReduce数据处理能力的一种尝试。
- HDFS：The Hadoop Distributed File System（HDFS）提供跨多个机器存储大型文件的一种解决方案。Hadoop 和 HDFS 都是从 Google File System（GFS）中派生的。Hadoop 2.0.0 之前，NameNode是HDFS集群的一个单点故障（SPOF）。利用ZooKeeper、HDFS高可用性特性解决了这个问题，提供选项来运行两个重复的NameNodes，在同一个集群中，同一个Active/Passive配置。
- HBase：灵感来源于Google BigTable。HBase是Google BigTable的开源实现，类似Google BigTable利用GFS作为其文件存储系统，HBase 利用 Hadoop HDFS作为其文件存储系统；Google运行MapReduce来处理BigTable中的海量数据，HBase同样利用 Hadoop MapReduce 来处理 HBase 中的海量数据；Google BigTable利用Chubby作为协同服务，HBase利用ZooKeeper 作为协同服务。
- Hive：Facebook开发的数据仓库基础设施，用于数据汇总、查询和分析。Hive 提供类似SQL的语言（不兼容 SQL92）——HiveQL。
- Pig：Pig提供一个引擎在 Hadoop 并行执行数据流。Pig 包含一种语言：Pig Latin，用来表达这些数据流。Pig Latin包括大量的传统数据操作（join、sort、filter等），也可以让用户开发他们自己的函数，用来查看、处理和编写数据。Pig在Hadoop 上运行，在Hadoop 分布式文件系统(HDFS)和Hadoop处理系统MapReduce中都可以使用。Pig使用MapReduce 来执行所有的数据处理，编译Pig Latin脚本，用户可以编写一个系列、一个或者多个的MapReduce作业，然后执行。Pig Latin看起来跟大多数编程语言都不一样，没有if状态和for循环。
- ZooKeeper：ZooKeeper是Hadoop的正式子项目，它是一个针对大型分布式系统的可靠协调系统，提供的功能包括：配置维护、名字服务、分布式同步、组服务等。ZooKeeper的目标就是封装好复杂且容易出错的关键服务，将简单易用的接口和性能高效、功能稳定的系统提供给用户。ZooKeeper是Google的Chubby一个开源的实现，是高效和可靠的协同工作系统。ZooKeeper能够用来执行Leader选举、配置信息维护等。在一个分布式的环境中，我们需要一个 Master 实例用来存储一些配置信息，确保文件写入的一致性等。
- Mahout：基于MapReduce的机器学习库和数学库。

1.3 Hadoop 3 新特性

由于Hadoop 2.0是基于JDK 1.7开发的，而JDK 1.7在2015年4月已停止更新，这直接迫使Hadoop社区基于JDK 1.8重新发布一个新的Hadoop版本,即Hadoop 3.0。Hadoop 3.0中引入了一些重要的功能和优化,包括HDFS可擦除编码、多NameNode支持、MR Native Task优化、YARN container resizing等。Hadoop 3.x以后将会调整方案架构,将MapReduce基于内存+IO+磁盘共同处理数据。改变最大的是HDFS,HDFS通过最近block块计算,根据最近计算原则,本地block块加入到内存先计算,再通过IO共享内存计算区域,最后快速形成计算结果。

根据官方change log(修改日志)的介绍,Hadoop 3新增的特性说明如下:

(1)最低支持JDK 1.8及以上版本。不再支持JDK 1.7。Hadoop版本与JDK版本之间的匹配关系为:

- Apache Hadoop 3.3及更高版本支持 Java 8和Java 11(仅运行时)。
- 请用Java 8编译Hadoop。不支持用Java 11编译Hadoop。
- Apache Hadoop从3.0.x到3.2.x目前只支持Java 8。
- Apache Hadoop从2.7.x到2.10.x支持Java 7和Java 8。

(2)YARN Timeline版本升为2.0。

(3)高可靠支持超过2个NameNode节点。如配置3个NameNode和5个JournalNode。

(4)默认端口变化,具体变化如图1-5所示。

分类	应用	Haddop 2.x port	Haddop 3 port
NNPorts	Namenode	8020	9820
NNPorts	NN HTTP UI	50070	9870
NNPorts	NN HTTPS UI	50470	9871
SNN ports	SNN HTTP	50091	9869
SNN ports	SNN HTTP UI	50090	9868
DN ports	DN IPC	50020	9867
DN ports	DN	50010	9866
DN ports	DN HTTP UI	50075	9864
DN ports	Namenode	50475	9865

图1-5　Hadoop 3的端口变化表

(5)从Hadoop 2.9开始添加了新的模块:Oozie,自此Hadoop拥有5个核心模块,以下是官方模块列表:

- Hadoop Common:支持其他Hadoop模块的常用工具。
- Hadoop分布式文件系统(HDFS):Hadoop用于数据存储的分布式文件系统,提供应用数据的高吞吐量访问。
- Hadoop YARN:用于作业调度和集群资源管理的框架。
- Hadoop MapReduce:基于YARN框架,用于大数据集的处理的分布式并行计算框架。
- Hadoop Ozone:是一个分布式对象存储系统,提供的是一个Key-Value形式的对象存储服务。

(6)Hadoop 3之后,已经不再建议使用root用户启动和管理Hadoop的进程。建议创建一个非root用户来启动和管理Hadoop的进程。建议创建一个名称为hadoop的用户,并设置hadoop用户属于wheel组。

1.4 虚拟机安装

本书将选择使用VirtualBox作为虚拟环境安装Linux和Hadoop。VirtualBox最早由SUN公司开发。由于SUN公司目前已经被Oracle收购，所以可以在Oracle公司的官方网站上下载到VirtualBox虚拟机软件的安装程序，下载地址为https://www.virtualbox.org。到笔者写作本书时，VirtualBox的最新版本为6.1.8。请访问VritualBox的官方网站，下载Windows Host版本的VirtualBox，其官方地址为https://www.virtualbox.org/wiki/Downloads，如图1-6所示。

同时，由于VitualBox需要虚拟化CPU的支持，如果在安装操作系统时，不支持虚拟化安装x64位的CentOS，可以在宿主机开机时按F12进入宿主机的BIOS设置，并打开CPU的虚拟化设置。

打开CPU的虚拟化设备，如图1-7所示。

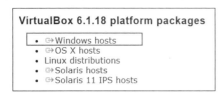

图1-6　VirtualBox下载地址

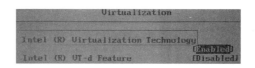

图1-7　CPU虚拟化设备

读者下载完成VirtualBox虚拟机后，自行安装即可。虚拟机的安装相对比较简单，以下是重要环节截图。

网络组件安装，直接单击"是"按钮，如图1-8所示。

网络组件安装，直接单击"安装"按钮，如图1-9所示。

图1-8　网络组件安装

图1-9　网络组件安装

网络组件安装成功后，会在操作系统的"我的网络"中多出一个Virtual Box Host Only的本地网卡，此网卡用于宿主机与虚拟机通信，如图1-10所示。

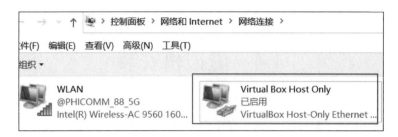

图1-10 本地网卡

1.5 安装Linux操作系统

本书将使用CentOS7作为环境来学习和安装Hadoop。首先需要下载CentOS操作系统，下载Minimal（最小）版本的即可，因为我们使用的CentOS并不需要可视化界面。CentOS官方网站为https://www.centos.org/，如图1-11所示。

Parent directory/	-	-
0_README.txt	2.4 KiB	2020-11-06 22:32
CentOS-7-x86_64-DVD-2009.iso	4.4 GiB	2020-11-04 19:37
CentOS-7-x86_64-DVD-2009.torrent	176.1 KiB	2020-11-06 22:44
CentOS-7-x86_64-Everything-2009.iso	9.5 GiB	2020-11-02 23:18
CentOS-7-x86_64-Everything-2009.torrent	380.6 KiB	2020-11-06 22:44
CentOS-7-x86_64-Minimal-2009.iso	973.0 MiB	2020-11-03 22:55
CentOS-7-x86_64-Minimal-2009.torrent	38.6 KiB	2020-11-06 22:44
CentOS-7-x86_64-NetInstall-2009.iso	575.0 MiB	2020-10-27 00:26
CentOS-7-x86_64-NetInstall-2009.torrent	23.0 KiB	2020-11-06 22:44
sha256sum.txt	398 B	2020-11-04 19:38
sha256sum.txt.asc	1.2 KiB	2020-11-06 22:37

图1-11 CentOS下载链接

下载完成以后，将得到一个CentOS-7-x86_64_Minimal-2009.iso文件。注意文件名中的2009不是指2009年，而是指2020年09月发布的版本。注意：清华大学镜像网站上也能下载到，而且速度很快。

接着启动VirtualBox，如图1-12所示。

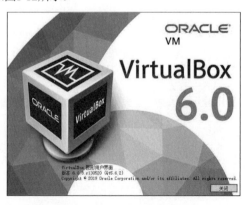

图1-12 VirtualBox启动界面

（1）在VirtualBox菜单上单击新建，打开向导，如图1-13所示。

（2）输入操作系统的名称和选择操作系统的版本，如图1-14所示。

图1-13　新建虚拟机　　　　　　　　　图1-14　选择将要安装的操作系统

（3）为新的系统分配内存，建议4GB（最少2GB）或以上，这要根据读者宿主机的内存而定。同时建议设置CPU为2个，如图1-15和图1-16所示。

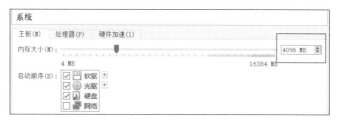

图1-15　设置内存大小

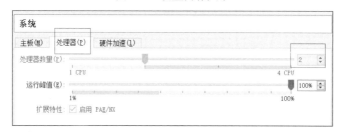

图1-16　设置处理器

（4）为新的系统创建硬盘，设置为动态增加，建议最大设置为30GB或以上。同时选择虚拟文件所保存的目录，默认的情况下，会将虚拟化文件保存到C:/盘上。笔者建议最好保存到非系统盘上，如D:/OS目录下是个不错的选择，如图1-17所示。

（5）选择创建以后，右击进入设置界面，在存储→盘片的位置选择已经下载好的CentOS7 ISO镜像文件，如图1-18所示。

（6）查看网络设置，将网卡1设置为NAT用于连接外网，将网卡2设置为Host-Only用于与宿主机进行通信。

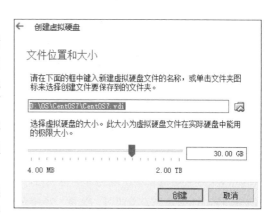

图1-17　选择保存目录

图1-18　选择镜像文件

网卡1的网络连接设置如图1-19所示。

网卡2的网络连接设置如图1-20所示。

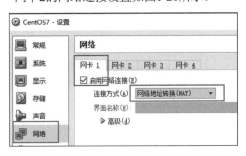 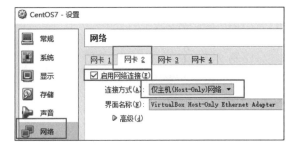

图1-19　设置网卡1网络连接　　　　　图1-20　设置网卡2网络连接

（7）现在启动这个虚拟机，将会进入安装CentOS Linux 7的界面，选择Install CentOS Linux 7，然后开始安装CentOS Linux，如图1-21所示。

图1-21　开始安装CentOS Linux

（8）在安装过程中出现选择语言项目，可以选择【中文】。选择安装位置，如图1-22所示。进入安装位置，选择整个磁盘即可，如图1-23所示。选择最小安装即可。注意，必须同时选择开启以太网络，如图1-24所示。否则安装成功以后，CentOS将没有网卡设置的选项。

图1-22　选择安装位置　　　　　　　　　图1-23　选择安装磁盘

图1-24　开启以太网

（9）在安装过程中，创建一个非root用户，并选择属于管理员组。在其后的操作中，笔者不建议使用root账户进行具体操作。一般情况下，非root用户只要执行sudo即可以用root用户执行相关命令，输入的密码并请牢记这个密码，如图1-25所示。

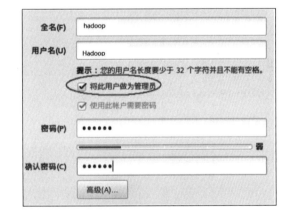

图1-25　创建密码

（10）在安装完成以后，重新启动，并测试是否可以使用之前创建的用户的账号和密码登录。刚开始安装完成后，请选择正常启动，正常启动即以有界面的方式启动，等我们设置好一些信息后，即可以选择无界面启动。

启动方式选择有界面启动，如图1-26所示。

（11）设置静态IP地址。启动后，将显示如图1-27所示的登录界面，此时可以选择以 root用户名和密码登录。注意输入密码时，将不会有任何响应，不必担心，只要确认输入正确，回车即可以看到登录成功后的界面，如图1-28所示。

对于Linux系统来说，如果当前用户是root用户，将会显示#，如图1-28所示。root用户登录成功后，将会显示 [root@server8 ~]#，其中#表示当前用户为root用户。如果是非root用户将显示为$。

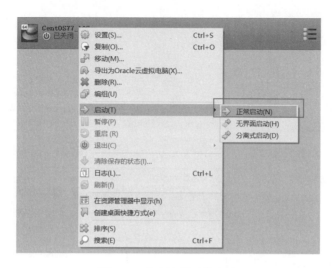

图1-26 右击选择正常启动

图1-27 登录界面

图1-28 登录成功后的界面

设置静态IP地址，使用vim修改/etc/sysconfig/network-scripts/ifcfg-enp0s8，修改内容如下：其中IPADDR=192.168.56.201为Linux的Host Only网卡地址，用于主机通信。输出完成以后，按ESC键，然后再输入：wq保存配置，退出即可。这是vim的基本操作，不了解的读者，可以去网上查看vim的基本使用。

```
TYPE=Ethernet
PROXY_METHOD=none
BROWSER_ONLY=no
BOOTPROTO=static
DEFROUTE=yes
IPV4_FAILURE_FATAL=no
IPV6INIT=yes
IPV6_AUTOCONF=yes
IPV6_DEFROUTE=yes
IPV6_FAILURE_FATAL=no
IPV6_ADDR_GEN_MODE=stable-privacy
NAME=enp0s8
UUID=620377da-1744-4268-b6d6-a519d27e01c6
DEVICE=enp0s8
ONBOOT=yes
IPADDR=192.168.56.201
```

请牢记上面设置的IP地址。现在可以关闭Linux系统，以"无界面启动"方式重新启动CentOS。以后我们将使用SSH客户端登录此CentOS。

上述文件是在配置了Host Only网卡的情况下，才会存在ifcfg-enp0s8文件。如果没有这个文件，请关闭Linux，并重新添加Host Only网卡后，再进行配置。如果添加了Host Only网卡后，依然没有此文件，可以在相同目录下，复制ifcfg-enp0s3为ifcfg-enp0s8创建此文件。

现在关闭CentOS，以无界面方式启动，如图1-29所示。

图1-29　以无界面方式启动

注意：

（1）本书不是讲VirtualBox虚拟机的使用，所以只给出具体的操作步骤。

（2）在安装过程中，鼠标会在虚拟机和宿主机之间切换。如果要从虚拟机中退出鼠标，直接按Ctrl键即可。

（3）关于Linux命令请读者自行参考Linux手册，如：vim/vi、sudo、ls、cp、mv、tar、chmod、chown、scp、ssh-keygen、ssh-copy-id、cat、mkdir等命令，将在后面经常使用到。

1.6　SSH工具与使用

读者可以选择xShell、CRT、MobaXterm等，作为Linux操作的远程命令行工具。同时配合它们提供的xFtp可以实现文件的上传与下载。

需要说明的是，xShell、CRT、MobaXterm上传的方式各不相同。读者可以查看各软件相关文档实现上传操作，如图1-30所示为MobaXterm的上传方式。

xShell和CRT是收费软件，不过，读者可以在安装时，选择 free for school（学校免费版本）免费使用。安装完成以后，在命令行使用SSH即可以登录Linux。MobaXterm为免费软件，图1-31所示为MobaXterm通过SSH远程登录Linux的界面。

图1-30　MobaXterm文件上传

图1-31　创建新的连接

在弹出的界面上选择SSH，并输入主机名称和用户名登录，如图1-32所示。

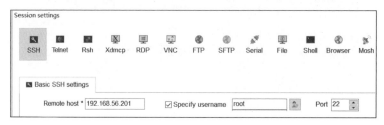

图1-32　输入主机名称并登录

输入密码不会有任何回显，只要输入正确，回车即可登录，如图1-33所示。

图1-33　输入密码

SSH登录成功以后的界面如图1-34所示。

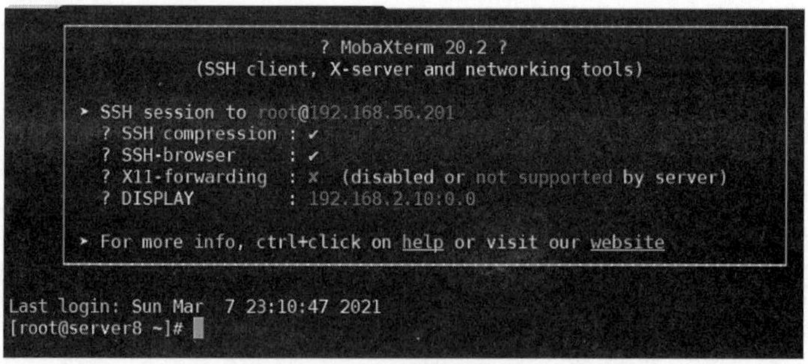

图1-34　SSH登录成功

1.7　Linux统一设置

本书后面的讲解将使用一些Linux统一的设置，在本节统一讲解。由于本次登录（见图1-29）是用root登录的，因此可以直接操作某些命令，不用再添加sudo命令。

1. 配置主机名称

笔者习惯将主机名取为server+IP最后一部分作为主机名称,所以取主机名为server201,因为本主机的IP地址设置为192.168.56.201。

```
# hostnamectl set-hostname server201
```

2. 修改hosts文件

在hosts文件的最后面,添加以下的配置,通过命令 vim /etc/hosts进行修改。

```
192.168.56.201 server201
```

3. 关闭和禁用防火墙

```
# systemctl stop firewalld
# systemctl disable firewalld
```

4. 禁用SELinux,需要重新启动

```
#vim /etc/selinux/config
SELINUX=disabled
```

5. 设置时间同步(可选)

```
#vim /etc/chrony.conf
```

删除所有server,只添加:

```
server ntp1.aliyun.com iburst
```

重新启动chronyd:

```
#systemctl restart chronyd
```

查看状态:

```
#chronyc sources -v
^* 120.25.115.20
```

如果显示*,则表示时间同步成功。

6. 在/usr/java目录下,安装JDK 1.8

usr目录为unix system resource目录,将JDK 1.8_x64安装到此目录下。

首先到Oracle官方网站下载JDK 1.8的Linux版本,如图1-35所示。

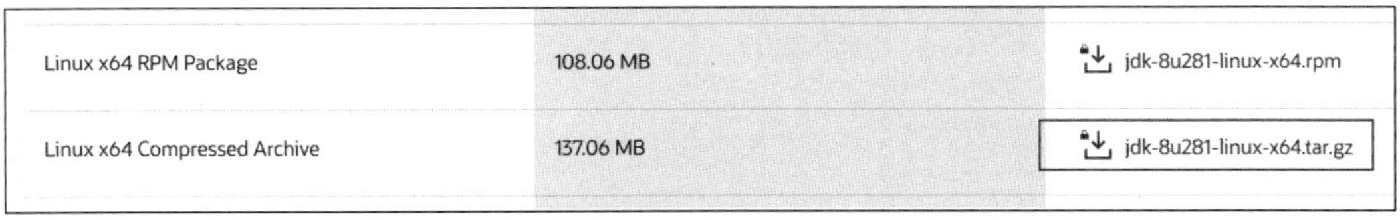

Linux x64 RPM Package	108.06 MB	jdk-8u281-linux-x64.rpm
Linux x64 Compressed Archive	137.06 MB	jdk-8u281-linux-x64.tar.gz

图1-35 JDK下载

JDK上传到Linux并解压:

```
# tar -zxvf jdk-8u281-linux-x64.tar.gz -C /usr/java/
```

7. 配置JAVA_HOME环境变量

```
#vim /etc/profile
```

在profile文件最后，添加以下配置：

```
export JAVA_HOME=/usr/java/jdk1.8.0_281
export PATH=.:$PATH:$JAVA_HOME/bin
```

让环境变量生效：

```
# source /etc/profile
```

检查Java版本：

```
[root@localhost bin]# java -version
java version "1.8.0_281"
Java(TM) SE Runtime Environment (build 1.8.0_192-b12)
Java HotSpot(TM) 64-Bit Server VM (build 25.192-b12, mixed mode)
```

到此，基本的Linux环境就已经配置完成了。

1.8 小　　结

本章主要讲解了以下的内容：

- 大数据概述。
- Hadoop介绍。
- 虚拟机及网络的配置。
- Linux操作系统CentOS7的安装过程。
- 使用xShell登录CentOS7。
- Linux的基本命令。
- SSH远程登录Linux。
- 配置通用的一些配置。
- 安装JDK及配置环境变量。

第 2 章

Hadoop伪分布式集群搭建

主要内容:

- ❖ 安装独立运行的Hadoop。
- ❖ Hadoop伪分布式的安装与配置。
- ❖ HDFS的命令。
- ❖ Java操作HDFS。

Hadoop的运行方式可以分为三种:

- 独立运行的Hadoop,不提供HDFS存储服务,也不需要启动任何的后台守护进程,但可以直接在本地运行MapReduce程序,并将输出结果保存到本地磁盘上。
- 伪分布式运行的Hadoop,一般是指只有一台服务器的Hadoop运行环境,需要启动NameNode(主节点存储服务)、SecondaryNameNode(主节点日志数据备份服务)可提供HDFS存储服务。启动守护进程ResourceManager和NodeManager运行MapReduce程序并将结果输出到HDFS上。
- 集群运行的Hadoop。可用于生产环境的高可用集群。借助ZooKeeper实现宕机容灾和自动切换。

为了快速上手Hadoop,我们会运行一个独立的MapReduce。独立运行的MapReduce可读取本地文本文件,然后将输出的数据保存到本地磁盘上。

注意:本书后面的环境,都将使用CentOS7、JDK 1.8_x64和Hadoop 3.2.2作为基础环境。

2.1 安装独立运行的Hadoop

独立运行的Hadoop可以帮助我们快速运行一个MapReduce官方示例,以了解MapReduce的运行方式。后面的测试和基本命令将会运行在分布式环境下。有些应用,如HBase、Hive需要真实的集群环境。

步骤 01 下载Hadoop。

Hadoop 3.2.2的下载地址为：https://www.apache.org/dyn/closer.cgi/hadoop/common/hadoop-3.2.2/hadoop-3.2.2.tar.gz。

步骤 02 解压并配置环境。

以hadoop用户登录，并在/home/hadoop的主目录下创建一个目录，用于安装Hadoop。

```
$ mkdir ~/program
```

上传Hadoop压缩包，并解压到program目录下：

```
$ tar -zxvf hadoop-3.2.2.tar.gz -C ~/program/
```

配置Java环境变量，修改Hadoop解压目录下的/etc/hadoop/hadoop-env.sh文件，找到${JAVA_HOME}并将其设置为本机JAVA_HOME的地址。

```
$ vim ~/program/hadoop-3.2.2/etc/hadoop/hadoop-env.sh
export JAVA_HOME=/usr/java/jdk1.8.0_281
```

配置Hadoop环境变量：

```
$ vim /home/hadoop/.bash_profile
export HADOOP_HOME=/home/hadoop/program/hadoop-3.2.2
export PATH=$PATH:$HADOOP_HOME/bin
```

注意：由于这里是用hadoop用户登录的，只配置了hadoop用户的环境变量，这种情况下，这种配置只会让当前用户可用。读者可以根据自己的要求进行配置。比如：如果配置到/etc/profile文件中，则是整个系统都可以使用环境变量，这种情况下，就不要将Hadoop安装到某个用户的主目录下了。

让环境变量生效：

```
$source ~/.bash_profile
```

输入hadoop命令，查看Hadoop的版本：

```
[hadoop@server201 ~]$ hadoop version
Hadoop 3.2.2
Source code repository Unknown -r 7a3bc90b05f257c8ace2f76d74264906f0f7a932
Compiled by hexiaoqiao on 2021-01-03T09:26Z
Compiled with protoc 2.5.0
From source with checksum 5a8f564f46624254b27f6a33126ff4
This command was run using
/home/hadoop/program/hadoop-3.2.2/share/hadoop/common/hadoop-common-3.2.2.jar
```

步骤 03 独立运行MapReduce。

Hadoop可以运行在一个非分布式的环境下，即可以运行为一个独立的Java进程。下面尝试运行一个WordCount的MapReduce示例。

创建一个任意的文本文件，并输入一行英文单词：

```
[hadoop@server201 ~]$ touch a.txt
[hadoop@server201 ~]$ vim a.txt
Hello This is
a Very Sample MapReduce
```

```
Example of Word Count
Hope You Run This Program Success!
```

执行WordCount测试：

```
[hadoop@server201 ~]$ hadoop jar \
~/program/hadoop-3.2.2/share/hadoop/mapreduce/hadoop-mapreduce-examples-3.2.2.jar \
 wordcount \
~/a.txt \
~/out
```

命令执行成功后，会显示以下信息，注意输出的日志会比较多，请仔细查找。

```
2021-03-08 21:59:19,536 INFO mapreduce.Job:  map 100% reduce 100%
2021-03-08 21:59:19,537 INFO mapreduce.Job: Job job_local215774179_0001 completed successfully
```

命令说明：

- hadoop jar用于执行一个MapReduce示例。在Linux中，如果命令有多行，可以通过输入"\"（斜线）换行。注意"\"前面必须有空格。
- hadoop-mapreduce-examples-3.2.2.jar为官方提供的示例程序包，WordCount是执行的任务，~/a.txt是输入的目录或文件，~/out是程序执行成功以后的输出目录。

程序执行成功以后，进入out输出目录，查看输出目录中的数据文件，其中part-r-0000为数据文件。_SUCCESS为标识成功的文件，里面没有数据。

```
[hadoop@server201 ~]$ cd out/
[hadoop@server201 out]$ ll
总用量 4
-rw-r--r-- 1 hadoop hadoop 122 3月   8 21:59 part-r-00000
-rw-r--r-- 1 hadoop hadoop   0 3月   8 21:59 _SUCCESS
```

通过cat查看part-r-00000文件中的数据，可以看到已经对a.txt中单词进行了数量统计，且默认排序为字母的顺序，字母后面跟的是此单词出现的次数。

```
[hadoop@server201 out]$ cat *
Count       1
Example     1
Hello       1
Hope        1
MapReduce   1
Program     1
Run         1
Sample      1
Success!    1
This        2
Very        1
Word        1
You         1
a           1
is          1
of          1
```

可见，已经对<input>目录中文件中的数据进行统计。至此，独立运行模式的Hadoop系统已经安装完成。

Hadoop独立运行方式只是一个练习环境，在正式生产环境中，不会使用这种方式。这里只是让读者了解一下MapReduce的运行。而且在此模式下，Hadoop的HDFS不会运行，也不会存储数据。

2.2 Hadoop伪分布式环境准备

Hadoop伪分布式，即在单机模式下运行Hadoop。我们需要运行5个守护进程，说明如下。

负责HDFS存储的进程三个（见图2-1）：

- NameNode进程：作为主节点，主要负责分配数据存储到位置。
- SecondaryNameNode进程：作为NameNode日志备份和恢复进程，避免数据丢失。
- DataNode进程：作为数据的存储节点，接收客户端的数据读写请求。

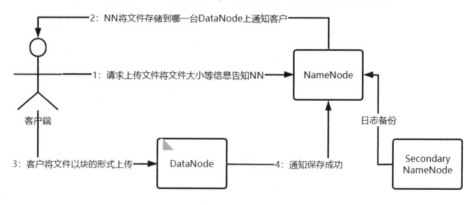

图2-1　HDFS存储进程

负责MapReduce计算的进程以下两个：

- ResourceManager进程：负责分配计算任务由哪一台主机执行。
- NodeManager进程：负责执行计算任务。

在真实集群环境下，部署的一般规则是：

- 由于NodeManager需要读取DataNode上的数据，用于执行计算，所以一般DataNode与NodeManager并存。
- 由于NameNode在运行时，需要在内存中大量缓存文件块的数据。因此，NameNode节点应该部署到内存比较大的主机上。
- 在真实的集群环境下，一般部署多个NameNode节点，相互之间互为备份和切换关系，且不再部署SecondaryNameNode进程。

伪分布式可以让读者快速学习HDFS的命令及开发MapReduce应用。对于学习Hadoop有很大的帮助。在安装之前，笔者有以下建议：

- 配置静态IP地址。虽然是单机模式，但也建议配置静态的IP地址，这样做有助于以后配置集群环境时固定IP，养成良好的习惯。
- 修改主机名称为一个便于记忆的名称，如server201，修改规则一般为本机的IP地址最后一段作为服务器的后缀，如192.168.56.201主机可以修改本主机的名称为server201。
- 由于启动Hadoop的各个进程使用的是SSH。所以，必须配置本机免密码登录。本章后面的步骤会讲到如何配置SSH免密码登录。配置SSH免密码登录的规则是在启动的集群的主机上，向其他主机配置SSH免密登录，以便于操作机可以在不登录其他主机的情况下，启动所需要的进程。
- 关闭防火墙。如果CentOS7没有安装防火墙，可以不用关闭；如果已经安装，请检查防火墙的状态，如果是运行状态请关闭防火墙并禁用防火墙。注意，在生产环境下，不要直接禁用防火墙，可以指定Hadoop的某些端口开放。
- 使用非root用户。前面章节我们创建了一个名为hadoop用户，此用户同时属于wheel组（拥有此组的用户可以使用sudo命令，执行一些root用户的操作）。我们就以此用户作为执行命令的用户。

步骤 01 配置静态IP地址。

前面的章节已经讲解静态IP的设置，此处再做详细讲解。使用SSH登录CentOS7。然后使用ifconfig查看IP地址，如果没有ifconfig命令，可以使用sudo yum -y install net-tools安装ifconfig命令。其实在CentOS7中，已经使用ip addr命令显示当前主机的IP地址。所以，也可以不安装net-tools。

```
$ ifconfig
enp0s3: flags=4163<UP,BROADCAST,RUNNING,MULTICAST>  mtu 1500
inet 10.0.2.15  netmask 255.255.255.0  broadcast 10.0.2.255
enp0s8: flags=4163<UP,BROADCAST,RUNNING,MULTICAST>  mtu 1500
inet 192.168.56.201  netmask 255.255.255.0
```

上例显示为两块网卡，其中enp0s3的IP地址为10.0.2.15，此网卡为NAT网络，用于上网。enp0s8的IP地址为192.168.56.201，此网卡为Host Only网络，用于与宿主机进行通信。我们要修改的就是enp0s8这个网卡，将它的IP地址设置为固定IP。

IP设置保存在文件中，这个文件为/etc/sysconfig/network-scripts/ifcfg-enp0s8。使用cd命令切换到这个目录下。使用ls命令显示这个目录下的所有文件，读者可能会发现只有ifcfg-enp0s3这个文件，现在使用cp命令将ifcfg-enp0s3复制一份并命名为ifcfg-enp0s8。由于etc目录不属于hadoop用户，所以操作时需要添加sudo前缀。

```
$ sudo cp ifcfg-enp0s3 ifcfg-enp0s8
```

使用vim命令修改为静态IP地址：

```
$ sudo vim ifcfg-enp0s8
```

将原来的dhcp修改成static，即静态的地址，并设置具体的IP地址。其中，每一个网卡都应该具有唯一的UUID，所以建议修改其上的任意一个值，以便与之前enp0s3的UUID不同。部分内容修改如下：

```
BOOTPROTO="static"
NAME="enp0s8"
UUID="d2a8bd92-cf0d-4471-8967-3c8aee78d101"
```

```
DEVICE="enp0s8"
IPADDR="192.168.56.201"
```

现在重新连接网络:

```
$ sudo systemctl restart network.service
```

重新连接网络后,再次查看,IP地址已经发生变化:

```
[hadoop@server201 ~]$ ifconfig
enp0s3: flags=4163<UP,BROADCAST,RUNNING,MULTICAST>  mtu 1500
        inet 10.0.2.15  netmask 255.255.255.0  broadcast 10.0.2.255
enp0s8: flags=4163<UP,BROADCAST,RUNNING,MULTICAST>  mtu 1500
        inet 192.168.56.201  netmask 255.255.255.0  broadcast 0x20<link>
lo: flags=73<UP,LOOPBACK,RUNNING>  mtu 65536
        inet 127.0.0.1  netmask 255.0.0.0
```

步骤02 修改主机名称。

使用hostname命令检查当前主机的名称:

```
$ hostname
localhost
```

使用hostnamectl命令修改主机名称:

```
$ sudo hostnamectl set-hostname server201
```

步骤03 配置hosts文件。

hosts文件是本地DNS解析文件。配置此文件可以根据主机名找到对应的IP地址。

使用vi命令打开这个文件,并在文件中追加以下配置:

```
$ sudo vim /etc/hosts
192.168.56.201 server201
```

步骤04 关闭防火墙。

CentOS7默认情况下,没有安装防火墙。可以通过sudo firewall-cmd --state命令检查防火墙的状态,如果显示command not found,一般为没有安装防火墙,此步可以忽略。使用以下命令检查防火墙的状态:

```
$ sudo firewall-cmd --state
running
```

running表示防火墙正在运行。以下命令用于停止和禁用防火墙:

```
$ sudo systemctl stop firewalld.service
$ sudo systemctl disable firewalld.service
```

步骤05 配置免密码登录。

配置免密码登录的主要目的就是在使用hadoop脚本启动Hadoop的守护进程时,不需要再提示用户输入密码。SSH免密码登录的主要实现机制就是在本地生成一个公钥,然后将公钥配置到需要被免密登录的主机上,登录时自己持有私钥与公钥进行匹配,如果匹配成功,则登录主机,否则登录失败。

可以使用ssh-keygen命令生成公钥和私钥文件,并将公钥文件复制到被SSH登录的主机上。使用ssh-keygen命令输入以后直接按两次回车键即可生成公钥和私钥文件:

```
[hadoop@server201 ~]$ ssh-keygen -t rsa
Generating public/private rsa key pair.
Enter file in which to save the key (/home/hadoop/.ssh/id_rsa):
Created directory '/home/hadoop/.ssh'.
Enter passphrase (empty for no passphrase):
Enter same passphrase again:
Your identification has been saved in /home/hadoop/.ssh/id_rsa.
Your public key has been saved in /home/hadoop/.ssh/id_rsa.pub.
The key fingerprint is:
SHA256:IDI032gBEDXhFVE1l6oYca5P4fkfIZRywyhgJ4Id/I4 hadoop@server201
The key's randomart image is:
+---[RSA 2048]----+
|=*%+*+..o ..     |
|.=oO.+.o +.      |
|  +.*+= *.       |
|   +ooo=..       |
|    o = +S .     |
|   E + = . .     |
|      o . .      |
|       . . .     |
|        ..       |
+----[SHA256]-----+
```

如上述提示信息得知,生成的公钥和私钥文件将被放到~/.ssh/目录下。其中id_rsa文件为私钥文件,rd_rsa.pub为公钥文件。现在我们再使用ssh-copy-id命令将公钥文件发送到目标主机。由于是登录本机,所以直接输入本机即可:

```
[hadoop@server201 ~]$ ssh-copy-id server201
/usr/bin/ssh-copy-id: INFO: Source of key(s) to be installed: "/home/hadoop/.ssh/id_rsa.pub"
The authenticity of host 'server201 (192.168.56.201)' can't be established.
ECDSA key fingerprint is SHA256:KqSRs/H1WxHrBF/tfM67PeiqqcRZuK4ooAr+xT5Z4OI.
ECDSA key fingerprint is MD5:05:04:dc:d4:ed:ed:68:1c:49:62:7f:1b:19:63:5d:8e.
Are you sure you want to continue connecting (yes/no)? yes 输入yes
/usr/bin/ssh-copy-id: INFO: attempting to log in with the new key(s), to filter out any that are already installed
/usr/bin/ssh-copy-id: INFO: 1 key(s) remain to be installed -- if you are prompted now it is to install the new keys
```

输入密码后按回车键,将会看到提示成功信息:

```
hadoop@server201's password:
Number of key(s) added: 1
Now try logging into the machine, with:   "ssh 'server201'"
and check to make sure that only the key(s) you wanted were added.
```

此命令执行以后,会在~/.ssh目录下多出一个用于认证的文件,其中保存了某个主机可以登录的公钥信息,这个文件为~/.ssh/authorized_keys。如果读者感兴趣,可以使用cat命令查看这个文件的具体内容。此文件中的内容就是id_rsa.pub文件中的内容。

现在再使用ssh server201命令登录本机,会发现不需要输入密码即可显示登录成功。

```
[hadoop@server201 ~]$ ssh server201
Last login: Tue Mar  9 20:52:56 2021 from 192.168.56.1
```

2.3 Hadoop 伪分布式安装

经过前面环境的设置,我们可以正式安装Hadoop伪分布式了。在安装之前,请确定是否安装了JDK 1.8,并正确配置了JAVA_HOME、PATH环境变量。

在磁盘根目录下创建一个app目录,并授权给hadoop用户。我们将会把Hadoop安装到此目录下。先切换到根目录下:

```
[hadoop@server201 ~]$ cd /
```

添加sudo前缀,使用mkdir创建/app目录:

```
[hadoop@server201 /]$ sudo mkdir /app
[sudo] hadoop 的密码
```

将此目录的所有权赋予hadoop用户和hadoop组:

```
[hadoop@server201 /]$ sudo chown hadoop:hadoop /app
```

切换进入/app目录:

```
[hadoop@server201 /]$ cd /app/
```

使用ll -d命令查看本目录的详细信息,可见此目录已经属于hadoop用户:

```
[hadoop@server201 app]$ ll -d
drwxr-xr-x 2 hadoop hadoop 6 3月  9 21:35 .
```

将Hadoop的压缩包上传到/app目录下,并解压。
使用ll查看本目录,可以看到目录中已经存在hadoop-3.2.2.tar.gz文件:

```
[hadoop@server201 app]$ ll
总用量 386184
-rw-rw-r-- 1 hadoop hadoop 395448622 3月  9 21:40 hadoop-3.2.2.tar.gz
```

使用tar -zxvf命令解压此文件:

```
[hadoop@server201 app]$ tar -zxvf hadoop-3.2.2.tar.gz
```

解压后,可以看到一个hadoop-3.2.2目录。
查看/app目录,发现hadoop-3.2.2目录已经存在。

```
[hadoop@server201 app]$ ll
总用量 386184
drwxr-xr-x 9 hadoop hadoop       149 1月  3 18:11 hadoop-3.2.2
-rw-rw-r-- 1 hadoop hadoop 395448622 3月  9 21:40 hadoop-3.2.2.tar.gz
```

删除hadoop-3.2.2.tar.gz文件,此文件已经不再需要。

```
[hadoop@server201 app]$ rm -rf hadoop-3.2.2.tar.gz
```

下面开始配置Hadoop。Hadoop的所有配置文件都在hadoop-3.2.2/etc/hadoop目录下。首先切换到此目录下,然后开始配置。

```
[hadoop@server201 hadoop-3.2.2]$ cd /app/hadoop-3.2.2/etc/hadoop/
```

在Hadoop官方网址上提供了关于伪分布式配置的完整教程，它的地址是https://hadoop.apache.org/docs/stable/hadoop-project-dist/hadoop-common/SingleCluster.html#Configuration。读者也可以根据此教程学习Hadoop伪分布式的配置。

步骤01 配置hadoop-env.sh文件。

hadoop-env.sh文件是Hadoop的环境文件，在此文件中需要配置JAVA_HOME变量。在此文件的第55行，输入以下配置，然后按ESC键，再输入:wq保存退出即可：

```
export JAVA_HOME=/usr/java/jdk1.8.0_281
```

步骤02 配置core-site.xml文件。

core-site.xml文件为HDFS的核心配置文件，用于配置HDFS的协议、端口号和地址。

注意Hadoop 3.0以后，HDFS的端口号建议使用8020端口，但如果查看Hadoop的官方网址示例，依然延续的是Hadoop 2之前的9000端口，以下配置我们将使用8020端口，只要保证配置的端口没有被占用即可。配置时注意大小写。

使用vim打开core-site.xml文件，进入编辑模式：

```
[hadoop@server201 hadoop]$ vim core-site.xml
```

在<configuration></configuration>两个标签之间输入以下内容：

```
<property>
    <name>fs.defaultFS</name>
    <value>hdfs://server201:8020</value>
</property>
<property>
<name>hadoop.tmp.dir</name>
    <value>/opt/datas/hadoop</value>
</property>
```

配置说明：

- fs.defaultFS：用于配置HDFS的主协议，默认为file:///。
- hadoop.tmp.dir：用于指定NameNode日志及数据的存储目录，默认为/tmp。

步骤03 配置hdfs-site.xml文件。

hdfs-site.xml文件用于配置HDFS的存储信息。使用vim打开hdfs-site.xml文件，并在<configuration></configuration>标签中输入以下内容：

```
<property>
    <name>dfs.replication</name>
    <value>1</value>
</property>
<property>
    <name>dfs.permissions.enabled</name>
    <value>false</value>
</property>
```

配置说明：

- **dfs.replication**：用于指定文件块的副本数量。HDFS特别适合于存储大文件，它会将大文件切分成每128MB一块，存储到不同的DataNode节点上，且默认将每一块备份2份，共3份，即此配置的默认值为3，最大为512MB。由于我们只有一个DataNode节点，所以这里将文件副本数量修改为1。
- **dfs.permissions.enabled**：访问时是否检查安全，默认为true。为了方便访问，暂时修改为false。

步骤04 配置mapred-site.xml文件。

mapred-site.xml文件用于配置MapReduce的配置文件。通过vim打开此文件，并在<configuration>标签中输入以下内容：

```
$ vim mapred-site.xml
<property>
   <name>mapreduce.framework.name</name>
   <value>yarn</value>
</property>
```

配置说明：

- **mapreduce.framework.name**：用于指定调试方式。这里指定使用YARN作为任务调用方式。

步骤05 配置yarn-site.xml文件。

由于上面指定了使用YARN作为任务调用方式，所以这里需要配置YARN的配置信息。同样，使用vim编辑yarn-site.xml文件，并在<configuration>标签中输入以下内容：

```
<property>
   <name>yarn.resourcemanager.hostname</name>
   <value>server201</value>
</property>
<property>
   <name>yarn.nodemanager.aux-services</name>
   <value>mapreduce_shuffle</value>
</property>
```

通过hadoop classpath命令获取所有classpath的目录，然后配置到上述文件中。

输入hadoop classpath命令由于尚没有配置Hadoop的环境变量，所以需要输入完整的Hadoop运行目录：

```
[hadoop@server201 hadoop]$ /app/hadoop-3.2.2/bin/hadoop classpath
```

输入完成后，将显示所有classpath信息：

```
/home/hadoop/program/hadoop-3.2.2/etc/hadoop:/home/hadoop/program/hadoop-3.2.2
/share/hadoop/common/lib/*:/home/hadoop/program/hadoop-3.2.2/share/hadoop/common/*
:/home/hadoop/program/hadoop-3.2.2/share/hadoop/hdfs:/home/hadoop/program/hadoop-3
.2.2/share/hadoop/hdfs/lib/*:/home/hadoop/program/hadoop-3.2.2/share/hadoop/hdfs/*
:/home/hadoop/program/hadoop-3.2.2/share/hadoop/mapreduce/lib/*:/home/hadoop/progr
am/hadoop-3.2.2/share/hadoop/mapreduce/*:/home/hadoop/program/hadoop-3.2.2/share/h
adoop/yarn:/home/hadoop/program/hadoop-3.2.2/share/hadoop/yarn/lib/*:/home/hadoop/
program/hadoop-3.2.2/share/hadoop/yarn/*
```

然后将上述的信息复制一下，并配置到yarn-site.xml文件中：

```
<property>
```

```xml
    <name>yarn.application.classpath</name>
    <value>
    /home/hadoop/program/hadoop-3.2.2/etc/hadoop:/home/hadoop/program/hadoop-3.2.2/share/hadoop/common/lib/*:/home/hadoop/program/hadoop-3.2.2/share/hadoop/common/*:/home/hadoop/program/hadoop-3.2.2/share/hadoop/hdfs:/home/hadoop/program/hadoop-3.2.2/share/hadoop/hdfs/lib/*:/home/hadoop/program/hadoop-3.2.2/share/hadoop/hdfs/*:/home/hadoop/program/hadoop-3.2.2/share/hadoop/mapreduce/lib/*:/home/hadoop/program/hadoop-3.2.2/share/hadoop/mapreduce/*:/home/hadoop/program/hadoop-3.2.2/share/hadoop/yarn:/home/hadoop/program/hadoop-3.2.2/share/hadoop/yarn/lib/*:/home/hadoop/program/hadoop-3.2.2/share/hadoop/yarn/*
    </value>
</property>
```

配置说明：

- yarn.resourcemanager.hostname：用于指定ResourceManager的运行主机，默认为0.0.0.0，即本机。
- yarn.nodemanager.aux-services：用于指定执行计算的方式为mapreduce_shuffle。
- yarn.application.classpath：用于指定运算时的类加载目录。

步骤06 配置workers文件。

workers文件之前的版本叫作slaves，但功能一样。主要用于在启动时同时启动DataNode和NodeManager。

编辑workers文件，并输入本地名称：

```
server201
```

步骤07 配置Hadoop环境变量。

编辑/etc/profile文件：

```
$ sudo vim /etc/profile
```

并在里面添加以下内容：

```
export HADOOP_HOME=/app/hadoop-3.2.2
export PATH=$PATH:$HADOOP_HOME/bin
```

使用source命令，让环境变量生效：

```
$ source /etc/profile
```

然后使用hdfs version命令查看环境变量是否生效，如果配置成功，则会显示Hadoop的版本：

```
[hadoop@server201 hadoop]$ hdfs version
Hadoop 3.2.2
Source code repository Unknown -r 7a3bc90b05f257c8ace2f76d74264906f0f7a932
Compiled by hexiaoqiao on 2021-01-03T09:26Z
Compiled with protoc 2.5.0
From source with checksum 5a8f564f46624254b27f6a33126ff4
This command was run using /app/hadoop-3.2.2/share/hadoop/common/hadoop-common-3.2.2.jar
```

步骤08 初始化Hadoop的文件系统。

在使用Hadoop之前，必须先初始化HDFS文件系统，初始化的文件系统将会在hadoop.tmp.dir配置的目录下生成，即上面配置的/app/datas/hadoop目录下。

```
$ hdfs namenode -format
```

在执行命令完成以后，如果在输出的日志中找到以下这句话，即为初始化成功：

```
Storage directory /opt/hadoop_tmp_dir/dfs/name has been successfully formatted.
```

步骤09 启动和停止HDFS和YARN。

启动和停止HDFS及YARN的脚本在$HADOOP_HOME/sbin目录下。其中start-dfs.sh为启动HDFS的脚本，start-yarn.sh为启动ResourceManager的脚本。以下分别启动HDFS和YARN：

```
[hadoop@server201 /]$ /app/hadoop-3.2.2/sbin/start-dfs.sh
[hadoop@server201 /]$ /app/hadoop-3.2.2/sbin/start-yarn.sh
```

启动完成以后，通过jps来查看Java进程快照，可以发现有六个进程正在运行：

```
[hadoop@server201 /]$ jps
12369 NodeManager
12247 ResourceManager
11704 NameNode
12025 SecondaryNameNode
12686 Jps
11839 DataNode
```

其中：NameNode、SecondaryNameNode、DataNode是通过start-dfs.sh脚本启动的。ResourceManager和NodeManager是通过start-yarn.sh脚本启动的。

在启动成功以后，也可以通过 http://server201:9870页面查看NameNode的信息，如图2-2所示。

图2-2　HDFS的Web界面

可以通过http://server201:8088查看MapReduce的信息，如图2-3所示。

步骤10 关闭HDFS和YARN。

关闭HDFS和YARN，执行stop-dfs.sh和stop-yarn.sh：

```
[hadoop@server201 /]$ /app/hadoop-3.2.2/sbin/stop-yarn.sh
Stopping nodemanagers
Stopping resourcemanager
[hadoop@server201 /]$ /app/hadoop-3.2.2/sbin/stop-dfs.sh
Stopping namenodes on [server201]
Stopping datanodes
Stopping secondary namenodes [server201]
```

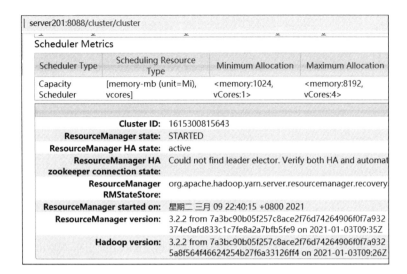

图2-3 MR Web界面

至此，Hadoop伪分布式模式安装配置成功。但是万里长征，我们才走了小小的一步。以下将讲解通过Hadoop的hdfs命令操作Hadoop的HDFS文件系统。

2.4 HDFS 操作命令

hdfs命令位于$HADOOP_HOME/bin目录下。由于已经配置了HADOOP_HOME和PATH的环境变量，所以此命令可以在任意目录下执行。可以直接在终端输入hdfs命令，查看它的使用帮助：

```
$ hdfs
Usage: hdfs [--config confdir] [--loglevel loglevel] COMMAND
       where COMMAND is one of:
dfs      run a filesystem command on the file systems supported in Hadoop.
classpath            prints the classpath
namenode -format     format the DFS filesystem
secondarynamenode    run the DFS secondary namenode
namenode             run the DFS namenode
journalnode          run the DFS journalnode
zkfc                 run the ZK Failover Controller daemon
datanode             run a DFS datanode
debug                run a Debug Admin to execute HDFS debug commands
dfsadmin             run a DFS admin client
haadmin              run a DFS HA admin client
fsck                 run a DFS filesystem checking utility
balancer             run a cluster balancing utility
jmxget               get JMX exported values from NameNode or DataNode.
mover                run a utility to move block replicas across
                     storage types
oiv                  apply the offline fsimage viewer to an fsimage
oiv_legacy           apply the offline fsimage viewer to an legacy fsimage
oev                  apply the offline edits viewer to an edits file
fetchdt              fetch a delegation token from the NameNode
```

```
getconf               get config values from configuration
groups                get the groups which users belong to
snapshotDiff          diff two snapshots of a directory or diff the
                      current directory contents with a snapshot
lsSnapshottableDir    list all snapshottable dirs owned by the current user
                       Use -help to see options
portmap               run a portmap service
nfs3                  run an NFS version 3 gateway
cacheadmin            configure the HDFS cache
crypto                configure HDFS encryption zones
storagepolicies       list/get/set block storage policies
version               print the version
Most commands print help when invoked w/o parameters.
```

上面的这些命令，在后面的讲解中基本都会涉及。现在让我们来查看几个使用比较多的命令。在上面的命令列表中，第一个dfs是经常使用到的命令，可以通过hdfs dfs -help查看dfs的具体使用方法。由于参数过多，本书就不一一列举了。dfs命令就是通过命令行操作HDFS目录或是文件的命令，类似于Linux文件命令一样，只不过操作的是HDFS文件系统中的文件。表2-1列出了几个常用命令。

表 2-1 HDFS 文件系统中的常用命令

命 令	功 能	示 例
-ls	用于显示 HDFS 文件系统上的所有目录和文件	hdfs dfs -ls /
-mkdir	在 HDFS 上创建一个新的目录	hdfs dfs -mkdir /test
-rm -r	删除 HDFS 上的一个目录，其中-r 参数为递归删除所有子目录。如果没有使用-r 参数，则是删除一个文件	hdfs dfs -rm -r /test
-put	将本地文件上传到 HDFS 中	hdfs dfs -put a.txt /test/a.txt
-cat	显示 HDFS 上某个文件中的所有数据，如果给出的是一个目录，则会忽略目录以下.（点）或是_（下画线）开头的文件	hdfs dfs -cat /test/a.txt
-get	从 HDFS 中将文件保存到本地	hdfs dfs -get /test/a.txt b.txt
-moveFromLocal	将本地文件上传到 HDFS 后并删除本地文件	hdfs dfs -moveFromLocal a.txt /test/a1.txt

示例：

显示根目录下的所有文件和目录：

```
[hadoop@server201 ~]$ hdfs dfs -ls /
Found 1 items
drwxr-xr-x   - hadoop supergroup          0 2021-03-10 20:41 /test
```

以递归的形式显示根目录下的所有文件和目录，注意-R参数：

```
[hadoop@server201 ~]$ hdfs dfs -ls -R /
drwxr-xr-x   - hadoop supergroup          0 2021-03-10 20:41 /test
-rw-r--r--   1 hadoop supergroup          6 2021-03-10 20:41 /test/a.txt
```

删除HDFS上的文件：

```
[hadoop@server201 ~]$ hdfs dfs -rm /test/a.txt
```

删除HDFS上的目录：

```
[hadoop@server201 ~]$ hdfs dfs -rm -r /test
```

将本地文件上传到HDFS：

```
[hadoop@server201 ~]$ hdfs dfs -copyFromLocal a.txt /test/a.txt
```

使用put命令，同样可以将本地文件上传到HDFS：

```
[hadoop@server201 ~]$ hdfs dfs -put a.txt /test/b.txt
```

使用moveFromLocal可以同时删除多个本地文件：

```
[hadoop@server201 ~]$ hdfs dfs -moveFromLocal a.txt /test/c.txt
```

使用get/copyToLocal/moveToLocal可以下载文件到本地：

```
[hadoop@server201 ~]$ hdfs dfs -get /test/c.txt a.txt
[hadoop@server201 ~]$ hdfs dfs -copyToLocal /test/a.txt a1.txt
[hadoop@server201 ~]$ hdfs dfs -moveToLocal /test/a.txt a2.txt
```

2.5　Java 项目访问 HDFS

我们不仅可以使用hdfs命令操作HDFS文件系统上的文件，还可以使用Java代码访问HDFS文件系统中的文件。

在Java代码中，操作HDFS主要通过以下几个主要的类：

- Configuration：用于配置HDFS。
- FileSystem：表示HDFS文件系统。
- Path：表示目录或是文件路径。

可以使用Eclipse或使用IDEA创建Java项目来操作HDFS文件系统。Eclipse是免费软件。IDEA有IC和IU两个版本，其中IU为IDEA Ultimate为完全功能版本，此版本需要付费后才能使用，我们可以选择免费使用的IC版本。后面在集成开发环境中将选择IDEA IC版本。

下载 IDEA IC 地址 为 https://www.jetbrains.com/idea/download/ #section=windows 。选择下载Community版本，如图2-4所示。选择下载zip文件即可，下载完成后，解压到任意目录下（建议没有中文没有空格的目录）。运行IdeaIC/bin目录下的idea64.exe即可以启动IDEA。以下我们开始创建Java项目，并通过Java代码访问HDFS文件系统。

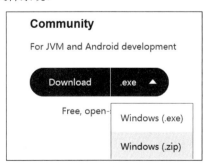

图2-4　下载IDEA IC版本

2.5.1 创建 Maven 项目

打开IDEA并选择创建新的项目，如图2-5所示。

图2-5 创建新项目

选择创建Maven项目，如图2-6所示。

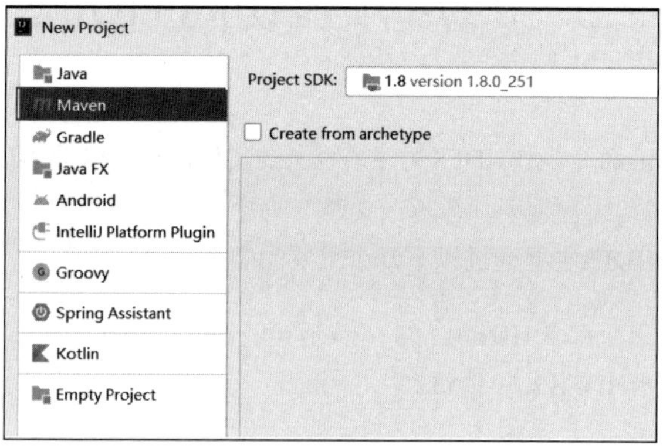

图2-6 创建Maven项目

选择项目创建的目录，并输入项目的名称，如图2-7所示。

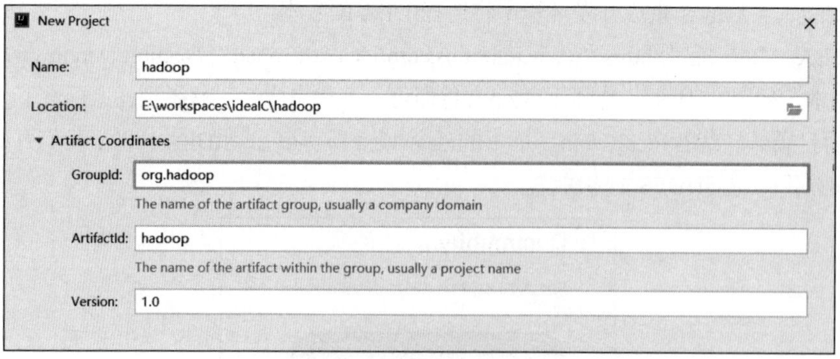

图2-7 输入项目名称

为了方便管理，我们以模块方式来开发，每一个章节可以为一个模块，而Hadoop是这些模块的父项目。

所以，在创建完成Hadoop项目后，修改Hadoop的项目类型为pom。以下代码是Hadoop父项目的pom.xml文件的部分内容，父项目的<package>类型为pom。

代码2.1　hadoop/pom.xml 文件

```xml
<groupId>org.hadoop</groupId>
<artifactId>hadoop</artifactId>
<version>1.0</version>
<packaging>pom</packaging>
```

在父项目中的dependencyManagement添加所需要的依赖后，子模块只需要添加依赖名称，不再需要导入依赖的版本。这样父项目就起到了统一管理版本的作用。

```xml
<dependencyManagement>
    <dependencies>
        <dependency>
            <groupId>org.apache.hadoop</groupId>
            <artifactId>hadoop-client</artifactId>
            <version>3.2.2</version>
        </dependency>
        <dependency>
            <groupId>junit</groupId>
            <artifactId>junit</artifactId>
            <version>4.13.2</version>
        </dependency>
    </dependencies>
</dependencyManagement>
```

现在我们再创建第一个模块。选择Hadoop项目，然后选择创建模块，如图2-8所示。

图2-8　创建子模块

输入模块的名称，如图2-9所示。

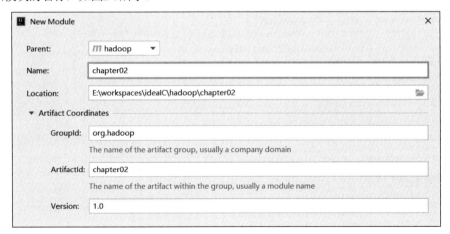

图2-9　输入模块名称

在创建的子模块chapter02中，修改pom.xml文件，添加以下依赖。注意，只输入groupId和artifactId，不需要输入版本，因为我们使用Hadoop项目作为父项目来管理依赖的版本。

```xml
<dependencies>
    <dependency>
        <groupId>org.apache.hadoop</groupId>
        <artifactId>hadoop-client</artifactId>
    </dependency>
    <dependency>
        <groupId>junit</groupId>
        <artifactId>junit</artifactId>
        <scope>test</scope>
    </dependency>
</dependencies>
```

查看 hadoop client Aggretator的依赖关系，如图2-10所示。

```
org.apache.hadoop:hadoop-client:3.2.2
    org.apache.hadoop:hadoop-common:3.2.2
    org.apache.hadoop:hadoop-hdfs-client:3.2.2
    org.apache.hadoop:hadoop-yarn-api:3.2.2
    org.apache.hadoop:hadoop-yarn-client:3.2.2
    org.apache.hadoop:hadoop-mapreduce-client-core:3.2.2
    org.apache.hadoop:hadoop-mapreduce-client-jobclient:3.2.2
    org.apache.hadoop:hadoop-annotations:3.2.2
```

图2-10 hadoop-client的依赖

至此，我们可以开发Java代码，并可以访问HDFS文件系统了。

2.5.2 HDFS 操作示例

1. 显示HDFS指定目录下的所有目录

使用fileSystem.listStatus方法，可以显示所有目录。

代码2.2 Demo01AccessHDFS.java

```java
01 System.setProperty("HADOOP_USER_NAME", "hadoop");
02 Configuration config = new Configuration();
03 config.set("fs.defaultFS", "hdfs://192.168.56.201:8020");
04 FileSystem fs = FileSystem.get(config);
05 FileStatus[] stas = fs.listStatus(new Path("/"));
06 for (FileStatus f : stas) {
07     System.out.println(f.getPermission().toString() + " "
08 + f.getPath().toString());
09 }
10 fs.close();
```

输出的结果如下所示。

```
rwxr-xr-x hdfs://192.168.56.201:8020/test
```

上面代码中，第01行用于设置访问Hadoop的用户名；第02行用于声明一个新的访问配置对象；第03行设置访问的具体地址；第04行创建一个文件系统对象；第05行~08行为输出根目录下的所有文件或目录，不包含子目录；第09行关闭文件系统。

2. 显示所有文件

可以使用fileSystem.listFiles函数显示所有文件,这个函数的第二个参数boolean用于指定是否递归显示所有文件。

代码2.3　Demo02ListFiles.java

```
01 System.setProperty("HADOOP_USER_NAME", "hadoop");
02 Configuration config = new Configuration();
03 config.set("fs.defaultFS", "hdfs://192.168.56.201:8020");
04 FileSystem fs = FileSystem.get(config);
05 RemoteIterator<LocatedFileStatus> files =
06 fs.listFiles(new Path("/"), true);
07 while(files.hasNext()){
08     LocatedFileStatus file = files.next();
09     System.out.println(file.getPermission()+" "+file.getPath());
10 }
11 fs.close();
```

添加了true参数以后,执行的结果如下所示。

```
rw-r--r-- hdfs://192.168.56.201:8020/test/a.txt
```

3. 读取HDFS文件的内容

读取HDFS文件的内容,可以使用fileSystem.open(...)打开一个文件输入流,然后就能读取文件流中的内容。

代码2.4　Demo03ReadFile.java

```
01 String server = "hdfs://192.168.56.201:8020";
02 System.setProperty("HADOOP_USER_NAME", "hadoop");
03 Configuration config = new Configuration();
04 config.set("fs.defaultFS", server);
05 try (FileSystem fs = FileSystem.get(config)) {
06     DataInputStream in = fs.open(new Path(server+"/test/a.txt"));
07     int len = 0;
08     byte[] bs = new byte[1024];
09     while((len=in.read(bs))!=-1){
10         String str = new String(bs,0,len);
11         System.out.print(str);
12 }}
```

4. 向HDFS中写入数据

向HDFS中写入数据,可以使用fileSystem.create/append方法获取一个OutputStream,然后向里面输入数据即可。

代码2.5　Demo04WriteFile.java

```
01 String server = "hdfs://192.168.56.201:8020";
02 System.setProperty("HADOOP_USER_NAME", "hadoop");
03 Configuration config = new Configuration();
```

```
04 config.set("fs.defaultFS", server);
05 try (FileSystem fs = FileSystem.get(config)) {
06     OutputStream out = fs.create(new Path(server+"/test/b.txt"));
07     out.write("Hello Hadoop\n".getBytes());
08     out.write("中文写入测试\n".getBytes());
09     out.close();
10 }
```

写入完成以后，通过cat查看文件中的内容。

```
[hadoop@server201 ~]$ hdfs dfs -cat /test/b.txt
Hello Hadoop
中文写入测试
```

其他方法如下所示，不再具体赘述：

- mkdirs：创建目录。
- create：创建文件。
- checkPath：创建一个文件检查点。
- delete：删除文件。

2.6　winutils

Hadoop通常运行在Linux上，而开发程序通常在Windows上，执行代码时，为了看到更多的日志信息，需要添加log4j.properties或log4j2的log4j2.xml。

通过查看hadoop-client-3.2.2依赖可知，系统中已经包含了log4j 1.2的日志组件，如图2-11所示。

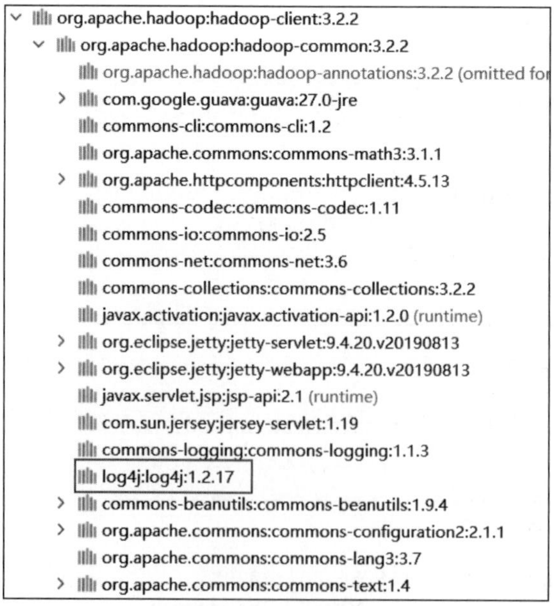

图2-11　log4j 1.2日志组件

此时只需要添加一个log4j与slf4j整合的依赖即可，所以添加以下依赖：

```xml
<dependency>
    <groupId>org.slf4j</groupId>
    <artifactId>slf4j-log4j12</artifactId>
    <version>1.7.26</version>
</dependency>
```

同时添加日志文件。直接在classpath下创建log4j.properties文件，即在项目的main/resources目录下添加log4j.properties文件即可：

```
log4j.rootLogger = debug,stdout
log4j.appender.stdout = org.apache.log4j.ConsoleAppender
log4j.appender.stdout.Target = System.out
log4j.appender.stdout.layout = org.apache.log4j.PatternLayout
log4j.appender.stdout.layout.ConversionPattern = [%-5p] %d{yyyy-MM-dd HH:mm:ss,SSS} method:%l%n%m%n
```

再次运行上面的访问HDFS的代码，将出现以下问题：

```
java.io.FileNotFoundException: HADOOP_HOME and hadoop.home.dir are unset
```

意思为HADOOP_HOME和hadoop.home.dir没有设置。

虽然出现上述问题，但程序仍然可以执行成功。出现这个问题的原因是本地没有Hadoop的程序。

此时，我们需要在本地，即Windows上解压一个相同的Hadoop版本，并配置HADOOP_HOME环境变量。比如笔者将Hadoop 3.2.2也同时解压到Windows系统的D:/program目录下，则配置环境变量为：

```
HADOOP_HOME=D:\program\hadoop-3.2.2
hadoop.home.dir=D:\program\hadoop-3.2.2
```

如果不想配置环境变量，也可以在上述的Java代码中添加这些变量到System中，如下所示。

```
System.setProperty("HADOOP_HOME", "D:/program/hadoop-3.2.2");
System.setProperty("hadoop.home.dir", "D:/program/hadoop-3.2.2");
```

同时还需要添加winutils，可以从GitHub或gitee网站上找到相关版本的winutils。将下载的文件放到D:\program\hadoop-3.2.2\bin中即可。再次运行，就没有任何警告信息了。从本书的配套资源中也可以找到相关资源，资源位置为chapter02/软件/winutils。

2.7 快速 MapReduce 程序示例

MapReduce为分布式计算模型，分布式计算最早由Google提出。MapReduce将运算的过程分为两个阶段，Map和Reduce阶段。用户只需要实现map和reduce两个函数即可。此处先为大家演示一个运行在本地的MapReduce程序，后续章节将会重点讲解MapReduce的开发方法。

之前我们讲过MapReduce可以直接在本地模式下运行。在项目中添加hadoop-mini-cluster依赖，即可直接在本地IDE环境中运行MapReduce程序，而不需要依赖于Hadoop集群环境，这在开发和测试工作中非常有用。为了帮助大家开发，我们在代码中注释了开发的步骤，具体代码的分析将会在MapReduce一章中详细讲解。

在本地运行Hadoop，需要将hadoop.dll文件放到windows/system32目录下，此文件可以在winutils目录下找到。

代码2.6　Demo02MapReduce.java

```java
01  package org.hadoop;
02  import org.apache.hadoop.conf.Configuration;
03  import org.apache.hadoop.conf.Configured;
04  import org.apache.hadoop.fs.FileSystem;
05  import org.apache.hadoop.fs.Path;
06  import org.apache.hadoop.io.LongWritable;
07  import org.apache.hadoop.io.Text;
08  import org.apache.hadoop.mapreduce.Job;
09  import org.apache.hadoop.mapreduce.Mapper;
10  import org.apache.hadoop.mapreduce.Reducer;
11  import org.apache.hadoop.mapreduce.lib.input.FileInputFormat;
12  import org.apache.hadoop.mapreduce.lib.input.TextInputFormat;
13  import org.apache.hadoop.mapreduce.lib.output.FileOutputFormat;
14  import org.apache.hadoop.mapreduce.lib.output.TextOutputFormat;
15  import org.apache.hadoop.util.Tool;
16  import org.apache.hadoop.util.ToolRunner;
17  import java.io.IOException;
18  /**
19   * MapReduce示例程序
20   */
21  //第1步：开发一个类，继承Configured实现接口Tool
22  public class Demo05MapReduce extends Configured implements Tool {
23      //第3步:添加main函数
24      public static void main(String[] args) throws Exception {
25          //第7步：开始任务
26          System.setProperty("HADOOP_HOME", "D:/program/hadoop-3.2.2");
27          System.setProperty("hadoop.home.dir", "D:/program/hadoop-3.2.2");
28          int res = ToolRunner.run(new Demo05MapReduce(),args);
29          System.exit(res);
30      }
31      //第2步：实现run函数
32      @Override
33      public int run(String[] strings) throws Exception {
34          //第6步：开发run函数内部代码
35          Configuration conf =getConf();
36          Job job = Job.getInstance(conf,"WordCount");
37          job.setJarByClass(getClass());
38          FileSystem fs = FileSystem.get(conf);
39          //声明输出目录
40          Path dest = new Path("D:/a/out");
41          //如果输出目录已存在，则删除
42          if(fs.exists(dest)){
43              fs.delete(dest,true);
44          }
45          //设置Mapper及Mapper输出的类型
46          job.setMapperClass(MyMapper.class);
47          job.setMapOutputKeyClass(Text.class);
48          job.setMapOutputValueClass(LongWritable.class);
49          //设置Reduce及Reduce的输出类型
50          job.setReducerClass(MyReduce.class);
```

```
51        job.setOutputKeyClass(Text.class);
52        job.setOutputValueClass(LongWritable.class);
53        //设置输入和输出类型
54        job.setInputFormatClass(TextInputFormat.class);
55        job.setOutputFormatClass(TextOutputFormat.class);
56        //设置输入输出目录
57        FileInputFormat.addInputPath(job,new Path("D:/a/a.txt"));
58        FileOutputFormat.setOutputPath(job,dest);
59        //开始执行任务
60        boolean boo = job.waitForCompletion(true);
61        return boo?0:1;
62    }
63    //第4步：开发Mapper类的实现类
64    public static class MyMapper extends
65 Mapper<LongWritable, Text,Text,LongWritable>{
66        @Override
67        protected void map(LongWritable key, Text value, Context context) throws
   IOException, InterruptedException {
68            Text outKey = new Text();
69            LongWritable outValue = new LongWritable(1L);
70            if(value.getLength()>0){
71                String[] strs = value.toString().split("\\s+");
72                for(String str:strs){
73                    outKey.set(str);
74                    context.write(outKey,outValue);
75                }
76            }
77        }
78    }
79    //第5步：开发Reduce程序
80    public static class MyReduce extends
   Reducer<Text,LongWritable,Text,LongWritable>{
81        @Override
82        protected void reduce(Text key, Iterable<LongWritable> values, Context
   context) throws IOException, InterruptedException {
83            long sum = 0;
84            LongWritable resultValue = new LongWritable(0);
85            for(LongWritable v:values){
86                sum+=v.get();
87            }
88            resultValue.set(sum);
89            context.write(key,resultValue);
90        }
91    }
92 }
```

运行后，查看D:\a\out目录下输出的文件：

```
._SUCCESS.crc
.part-r-00000.crc
_SUCCESS
part-r-00000
```

打开part-r-0000即为字符统计的结果，根据源文件不同，统计的结果也会不同，以下仅为参考。默认会以Key排序，所以输出的数据是以字母作为顺序排序输出的。

```
Configuration     1
Configured     1
Context     2
Demo05MapReduce     1
Demo05MapReduce(),args);     1
Exception     2
```

2.8 小　　结

本章主要讲解了以下内容：

- Hadoop伪分布式的环境。
- Hadoop的配置文件。
- 使用Hadoop的脚本start-dfs.sh和start-yarn.sh启动HDFS和YARN。
- Hadoop启动以后的六个进程。
- Hadoop启动以后，通过9870、8088两个端口访问HDFS和MapReduce的Web界面。
- HDFS的命令行操作。
- Java操作HDFS。
- 第一个运行在本地的MapReduce程序示例。

第 3 章

HDFS分布式存储实战

主要内容：

- ❖ HDFS的体系结构。
- ❖ HDFS命令。
- ❖ RPC远程调用。
- ❖ Hadoop各进程的功能。

3.1 HDFS的体系结构

本章将详细讲解HDFS的组件及其之间的关系。图3-1展示了客户端如何与HDFS进行交互并存取数据的过程。

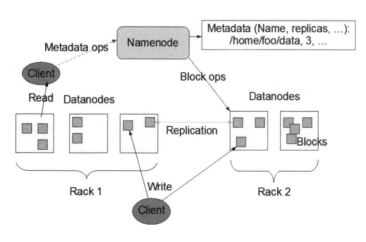

图3-1 HDFS的体系结构（图片引用自Hadoop官网）

图3-1的具体含义说明如下：

（1）客户端与NameNode进行交互，NameNode通知Client将数据保存在哪一台DataNode上。由Client保存数据到其中一个DataNode上，然后DataNode会根据元数据信息，再将自己的数据备份到其他DataNode上。这样，Client只要上传到其中一个DataNode上，DataNode之间负责数据副本的复制。NameNode中metadata用于在内存保存数据保存的元信息。在${hadoop.tmp.dir}/dfs/name目录下，保存了edits和fsimages文件，它们分别是日志文件和metadata的内存镜像文件。

（2）在保存成功以后，由NameNode来保存数据的元信息。即什么数据保存到哪几台DataNode节点上。

（3）每一个文件块的大小在Hadoop 2.0以后为128MB。可在hdfs-site.xml中配置dfs.blocksize，默认值为134 217 728字节，即128MB。文件块可以理解为将文件分割和存储的大小，如一个文件为138MB大小，此文件就会分为两个文件块即128MB一块和10MB一块，每一块都可以保存到不同的DataNode上。在获取文件中有多个文件块时，会再合并成一个完整的文件。同时，按HDFS每一个文件块存在3个副本的原则，则138MB的文件保存到HDFS文件系统上后，大小为128MB*3份+10MB*3份。而在NameNode中则保存了这六个文件块的具体位置信息，即元数据信息。

（4）SecondaryNameNode用于时时管理NameNode中的元数据信息。执行更新和合并的工作。

3.2 NameNode的工作

在伪分布式环境中，NameNode只有一个。但在分布式环境中，NameNode被抽象为NameService。而每一个NameService下最多可以有两个NameNode（Hadoop 3以后可以有多个）。这种情况下，一个NameNode为Active（活动），另一个为Standby（备份）。NameNode主要用于保存元数据、接收客户端的请求。NameNode的具体工作为：

（1）保存metadate元数据，始终在内存中保存metadata元数据信息用于处理读请求。由于每一个文件块都带有一个元数据信息，每个文件的NameNode元数据信息大概是250Byte，4000万个小文件的元数据信息就是4000万*250Byte≈10GB，也就是说，NameNode就需要10GB内存来存储和管理这些小文件的元数据信息。如果存储一个128MB的大文件，则只需要一个元数据信息，此时NameNode的内存只占用250Byte。如果存储128个小文件，每一个文件为1MB，则此时会有128个元数据信息，虽然存储量相同，但NameNode管理这些文件块所用的内存并不相同，后者将使用128*250Byte的内存大小来保存这些小文件的元数据信息。所以HDFS文件系统擅长处理大文件的场景，不擅长处理小文件。

（2）维护fsimage文件，也就是metadate元数据信息的镜像文件。此文件保存在$hadoop.tmp.dir所配置的目录中的name目录下。以下是我们看到的fsimage文件所位于的目录及目录下的文件。

```
[hadoop@server201 current]$ pwd
/app/datas/hadoop/dfs/name/current
[hadoop@server201 current]$ ll
总用量 3116
-rw-rw-r-- 1 hadoop hadoop      42 3月   9 22:40
edits_0000000000000000001-0000000000000000002
```

```
    -rw-rw-r-- 1 hadoop hadoop 1048576 3月  11 22:06
edits_inprogress_0000000000000000027
    -rw-rw-r-- 1 hadoop hadoop     630 3月  11 22:06 fsimage_0000000000000000026
    -rw-rw-r-- 1 hadoop hadoop      62 3月  11 22:06 fsimage_0000000000000000026.md5
    -rw-rw-r-- 1 hadoop hadoop       3 3月  11 22:06 seen_txid
    -rw-rw-r-- 1 hadoop hadoop     218 3月  11 20:06 VERSION
```

（3）当写请示到来时，首先写editlog即向edits写日志，成功返回以后再写入内存。

（4）SecondaryNameNode同时维护fsimage和edits文件，以更新NameNode的metadata元数据。

图3-2展示了NameNode是如何把元数据保存到磁盘上的。这里有两个不同的文件：

（1）fsimage文件：它是在NameNode启动时对整个文件系统的快照，是metadata的镜像。

（2）edit logs文件：每当写操作发生时，NameNode会首先修改这个文件，然后再去修改metadata。

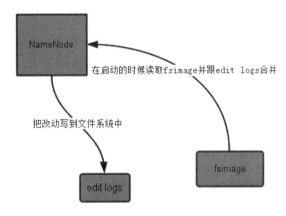

图3-2　将元数据保存到磁盘

元数据信息包含：

（1）fsimage为元数据的镜像文件，用于保存一段时间NameNode中元数据的信息。

（2）edits保存了数据的操作日志。

（3）fstime保存最近一次checkpoint的时间。fstime保存在内存中。

这些文件都保存在dfs.namenode.edits.dir配置的目录下，如果没有配置此项，则默认保存到file://${hadoop.tmp.dir}/dfs/name目录下，正如在布伪分布式环境中我们配置了hadoop.tmp.dir=/app/datas/hadoop，则这些文件都保存到/app/datas/hadoop/dfs/name/current目录下。

NameNode始终在内存中保存metadate元数据信息。在处理读写数据时，会先写edits到磁盘，成功返回以后再修改内存中的metadata元数据。

NameNode会维护一个fsimage文件，此文件是metadata保存在磁盘上的镜像文件（Hadoop 2/3中fsimage与metadata操作实时同步。Hadoop 1不是实时同步）。每隔一段时间，SecondaryNameNode会合并fsimage和edits来更新内存中的matedata。

3.2.1　查看镜像文件

查看fsimage文件内容可以通过hdfs oiv命令，此命令用于将fsimage文件转换成可读的xml文件。oiv的具体含义，可以通过help命令获取：

```
[hadoop@server201 current]$ hdfs oiv -h
Usage: bin/hdfs oiv [OPTIONS] -i INPUTFILE -o OUTPUTFILE
Offline Image Viewer
View a Hadoop fsimage INPUTFILE using the specified PROCESSOR,
saving the results in OUTPUTFILE.
```

通过上述命令结果可以看到oiv的含义为Offline Image View，即离线fsimage文件查看器。oiv常用参数为：

```
Required command line arguments:
-i,--inputFile <arg>   fsimage or XML file to process.

Optional command line arguments:
-o,--outputFile <arg>  Name of output file. If the specified
                       file exists, it will be overwritten.
                       (output to stdout by default)
                       If the input file was an XML file, we
                       will also create an <outputFile>.md5 file.
-p,--processor <arg>   Select which type of processor to apply
                       against image file. (XML|FileDistribution|
                       ReverseXML|Web|Delimited)
                       The default is Web.
```

参数说明：

- -i,--inputFile：用于指定fsimage文件或XML文件名称。
- -o,--outputFile：输出文件名称，如果文件已经存在则会覆盖，同时会生成一个*.md5比对文件。
- -p,--processor：处理方式，默认为Web，建议使用XML。

现在我们使用oiv命令，查看一个已经存在的fsimage文件，其格式如下：

```
[hadoop@current]$ hdfs oiv -i fsimage_0000000000000000024 -o ~/fsimage.xml -p XML
```

上面的命令，用于将fsimage文件转成XML文件，转换成功后，就可以使用vim查看fsimage文件的内容了。打开此文件，并格式化后的内容如下（根据不同的环境内容会有不同，仅作参考）：

```xml
<?xml version="1.0" encoding="utf-8"?>
<fsimage>
    <version>
        <layoutVersion>-65</layoutVersion>
        <onDiskVersion>1</onDiskVersion>
        <oivRevision>7a3bc90b05f257c8ace2f76d74264906f0f7a932</oivRevision>
    </version>
    ...
</fsimage>
```

可见在fsimage文件中，保存了版本、文件、节点等信息。

3.2.2 查看日志文件

hdfs oev命令用于查看edits日志文件，同样可以通过help命令，查看它的具体使用方法：

```
[hadoop@server201 current]$ hdfs oev -h
Usage: bin/hdfs oev [OPTIONS] -i INPUT_FILE -o OUTPUT_FILE
```

```
Offline edits viewer
Parse a Hadoop edits log file INPUT_FILE and save results
in OUTPUT_FILE.
```

可见oev的含义为offline edits viewer，即离线edits文件查看器。现在我们使用此命令，将edits文件转成可读的XML文件：

```
$ hdfs oev -i edits_0000000000000000013-0000000000000000014 -o ~/edits.xml -p XML
```

查看得到的XML文件内容如下，可见此文件中只记录了操作的事务日志信息。

```
<?xml version="1.0" encoding="UTF-8" standalone="yes"?>
<EDITS>
    <EDITS_VERSION>-65</EDITS_VERSION>
    <RECORD>
        <OPCODE>OP_START_LOG_SEGMENT</OPCODE>
        <DATA>
            <TXID>13</TXID>
        </DATA>
    </RECORD>
    <RECORD>
        <OPCODE>OP_END_LOG_SEGMENT</OPCODE>
        <DATA>
            <TXID>14</TXID>
        </DATA>
    </RECORD>
</EDITS>
```

3.2.3 日志文件和镜像文件的操作过程

在edits文件中，有一个名称为edits_inprogress_开头的文件，表示正在操作的日志文件。如当前目录下，我们正好有一个这样的日志文件，它的名称为：

```
edits_inprogress_0000000000000000028
```

此文件，将时时记录我们对HDFS文件的操作，如现在删除一个文件：

```
$ hdfs dfs -rm /test/a.txt
```

然后通过oev命令，将edits_inprocess_开始的文件转成XML文件，并读取里面的内容。

```
$ hdfs oev -i edits_inprogress_0000000000000000028 -o ~/inprocess.xml -p XML
```

我们会在文件的最后发现删除文件的操作记录：

```
<RECORD>
    <OPCODE>OP_DELETE</OPCODE>
    <DATA>
        <TXID>30</TXID>
        <LENGTH>0</LENGTH>
        <PATH>/test/a.txt</PATH>
        <TIMESTAMP>1615559516088</TIMESTAMP>
        <RPC_CLIENTID>68c4f333-8bad-472a-8990-4551ce0e5c2c</RPC_CLIENTID>
        <RPC_CALLID>3</RPC_CALLID>
    </DATA>
</RECORD>
```

其中OPCODE的值为OP_DELETE，即为删除文件，在DATA节点中则记录了文件的详细信息。由此可见，edits文件会时时记录文件的操作日志。

合并fsimage镜像文件

默认情况下，关闭和打开HDFS进程，则会自动合并并保存fsimage文件到磁盘。不过，也可以通过图3-3中所示的命令，将内存中的matedata保存为新的fsimage文件，首先查看目前的日志文件，已知最新的版本为28，即edits_inprogress_xxx28这个数字，就是最新的日志版本。而fsimage即内存中的matedata保存到磁盘上镜像文件总是比值小一个版本，即27。

```
     42 3月    9 22:40 edits_0000000000000000001-0000000000000000002
1048576 3月    9 22:40 edits_0000000000000000003-0000000000000000003
    611 3月   10 21:35 edits_0000000000000000004-0000000000000000012
     42 3月   10 22:35 edits_0000000000000000013-0000000000000000014
1048576 3月   10 22:42 edits_0000000000000000015-0000000000000000021
     88 3月   11 21:06 edits_0000000000000000022-0000000000000000026
     42 3月   11 22:06 edits_0000000000000000025-0000000000000000026
1048576 3月   11 22:06 edits_0000000000000000027-0000000000000000027
1048576 3月   12 22:31 edits_inprogress_0000000000000000028
    630 3月   11 22:06 fsimage_0000000000000000026
     62 3月   11 22:06 fsimage_0000000000000000026.md5
    630 3月   12 22:27 fsimage_0000000000000000027
     62 3月   12 22:27 fsimage_0000000000000000027.md5
      3 3月   12 22:27 seen_txid
    218 3月   12 22:27 VERSION
```

图3-3　合并fsimage镜像文件

通过输出seen_txid可知，如果下一次合并，则版本号为32。

```
[hadoop@server201 current]$ cat seen_txid
32
```

上面结果说明：fsimage_0000000000000000027还没有与edits_inprogress_0000000000000000028合并。现在我们使用以下操作执行合并工作。

先进入安全模式，通过dfsadmin可以使HDFS进入安全模式，进入安全模式后，将不能操作HDFS文件系统中的文件，一般在HDFS刚刚启动时，会有一段时间为安全模式，等所有的fsimage文件加载到内存后，就会自动退出安全模式。

使用dfsadmin命令让HDFS进入安全模式：

```
[hadoop@server201 current]$ hdfs dfsadmin -safemode enter
Safe mode is ON
```

也可以通过dfsadmin get查看当前是否处于安全模式：

```
[hadoop@server201 current]$ hdfs dfsadmin -safemode get
Safe mode is ON
```

保存matedata数据到fsimage中：

```
[hadoop@server201 current]$ hdfs dfsadmin -saveNamespace
Save namespace successful
```

退出安全模式：

```
[hadoop@server201 current]$ hdfs dfsadmin -safemode leave
Safe mode is OFF
```

现在再来查看日志文件,已经发生变化,如图3-4所示。说明已经将内存中的元数据保存到新的fsimage镜像文件中了。

```
 9 22:40 edits_0000000000000000001-0000000000000000002
 9 22:40 edits_0000000000000000003-0000000000000000003
10 21:35 edits_0000000000000000004-0000000000000000012
10 22:35 edits_0000000000000000013-0000000000000000014
10 22:42 edits_0000000000000000015-0000000000000000021
11 21:06 edits_0000000000000000022-0000000000000000024
11 22:06 edits_0000000000000000025-0000000000000000026
11 22:06 edits_0000000000000000027-0000000000000000027
12 22:47 edits_0000000000000000028-0000000000000000031
12 23:08 edits_0000000000000000032-0000000000000000033
12 23:08 edits_inprogress_0000000000000000034
12 22:27 fsimage_0000000000000000027
12 22:27 fsimage_0000000000000000027.md5
12 23:08 fsimage_0000000000000000033
12 23:08 fsimage_0000000000000000033.md5
12 23:08 seen_txid
12 23:08 VERSION
```

图3-4 执行安全模式后的日志文件结果

3.3 SecondaryNameNode

SecondaryNameNode为HA(高可用)集群的一个解决方案。在伪分布式中一般只有一个NameNode,一旦NameNode宕机,内存中的元数据信息也将随之丢失,而NameNode中的元数据信息,就像是书中的目录一样,记录了所有文件存储的信息。如果再次启动NameNode不能找回元数据信息,也将会无法找到之前存储的文件。而此时SecondaryNameNode就扮演了非常重要的角色,它时时将NameNode的元数据保存到磁盘上,NameNode重新启动时,会将磁盘上的元数据信息读取到内存,从而将损失降到最低。SecondaryNameNode的职责是合并NameNode的edit logs到fsimage文件中,如图3-5所示。在正式的集群中没有SecondaryNameNode进程,而是多个NameNode进程互为主备。

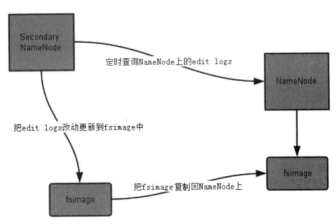

图3-5 SecondaryNameNode的工作过程

图3-5展示了SecondaryNameNode的工作过程,具体解释如下:

(1)它定时到NameNode去获取edit logs(内存中的edit logs即为metadata),并更新到fsimage上。具体执行合并更新的时间,可以通过以下两个参数在hdfs-site.xml文件中进行配置,以最先达到的值为执行的时间点。

- 时间参数：dfs.namenode.checkpoint.period，默认值为3600s，即3600秒。
- 事务参数：dfs.namenode.checkpoint.txns，默认值为1000000，即100万事务后将执行合并。

（2）NameNode在下次重启时会使用这个新的fsimage文件和edits文件进行合并，将合并到内存中的fsimage的元数据信息重新保存到磁盘上。

（3）SecondaryNameNode的整个目的是在HDFS中提供一个检查点。它只是NameNode的一个助手。

现在，我们明白了SecondaryNameNode所做的事情不过是帮助NameNode更好地进行工作。它不是要取代NameNode，也不是NameNode的备份。

3.4 DataNode

DataNode作为数据存储的节点，并时时与NameNode通过心跳机制保持通信。
DataNode的功能如下：

（1）提供真实的存储服务。

（2）在Hadoop 2.0及以后的版本中，每一个文件块的大小为128MB，文件块默认大小为128MB，如果一个文件没有128MB，则上传的文件将会占用一个实际大小的空间；如果文件大于128MB，则文件将会被分割成多个文件块。我们可以通过上传一个大于128MB的文件，而后查看上传以后文件是否分成多个文件的形式保存来验证。文件的块大小可以在hdfs-site.xml文件中添加配置dfs.blocksize，默认值为134217728，即128MB。

（3）如果在core-site.xml中配置了hadoop.tmp.dir，真实的数据将保存在${hadoop.tmp.dir}/data目录下；如果没有配置，则默认会将数据保存到/tmp/hadoop-${user.name}。在上述的目录中，有一个data目录，就是保存HDFS真实数据的位置。

（4）默认的副本为3个，在hdfs-site.xml中配置dfs.replication可以修改默认副本数量，最大值为512MB，默认值为3。

3.5 HDFS 的命令

在Hadoop 1.x版本中，使用hadoop命令管理HDFS文件系统。在Hadoop 2.x版本中，使用hdfs命令管理HDFS文件系统。

以下是Hadoop 1.x版本的命令，现在依然可以使用：

```
# hadoop fs -ls /
Found 2 items
drwxr-xr-x   - root supergroup          0 2018-12-09 21:45 /test
drwx------   - root supergroup          0 2018-12-09 20:46 /tmp
```

也可以省去hdfs://server201:8020，直接输入/（斜线）：

```
[root@server51 ~]# hdfs dfs -ls /
Found 2 items
drwxr-xr-x   - root supergroup          0 2018-12-09 21:45 /test
drwx------   - root supergroup          0 2018-12-09 20:46 /tmp
```

以下介绍几个常用的命令。

(1) 显示服务器文件列表：

```
hdfs dfs -ls /
```

(2) 将本地文件复制到HDFS上去：

```
$hdfs dfs -copyFromLocal  ~/home/wangjian/some.txt  /some.txt
```

(3) 查看服务器上的文件内容：

```
$hdfs dfs -cat /some.txt
```

(4) 从服务器上下载文件到本地：

```
$hdfs dfs -copyToLocal /test1.txt test1.txt
```

(5) 服务器文件和文件夹数：

```
$hdfs dfs -count /
```

(6) 向服务器上传文件：

```
$hdfs dfs -put test1.txt /test2.txt
```

(7) 从服务器获取文件到本地：

```
$hdfs dfs -get /test2.txt test3.txt
```

3.6 远程过程调用

在Hadoop中，多个进程（例如NameNode、DataNode等）之间数据的传递和访问都是通过RPC实现的。RPC是不同进程之间方法调用的解决方案。RPC调用的原理是通过网络代理实现远程方法的调用。这些功能已经被Hadoop封装，直接使用Hadoop提供的类，即可实现RPC远程过程调用。

由于被调用的类是通过动态代理实现的，因此必须拥有一个接口，而且接口上必须拥有一个public static final long versionID=xxxxL的声明，即序列化的versionID字段。

下面通过调用Hadoop RPC代码示例，简单演示RPC在多个进程之间的调用过程。

步骤01 开发一个接口。

在接口中，必须拥有一个versionID字段用于标识当前接口。

代码3.1 IHello.java

```
01 package org.hadoop.rpc.service;
02 public interface IHello {
03     //必须声明一个versionID值是任意的Long值
```

```
04     long versionID = 937530L;
05     //定义一个方法,用于远程测试调用
06     String say(String name);
07 }
```

步骤02 开发实现类,并实现接口中的方法。

实现接口,并实现接口中的方法。声明一个类,并声明一个方法,返回一个任意的字符串,一个类如果希望被远程RPC调用,这个类必须实现一个接口,因为内部的原理是JDK动态代理。

代码 3.2　HelloImpl.java

```
01 package org.hadoop.rpc.service;
02 import java.time.LocalDateTime;
03 public class HelloImpl implements IHello{
04     @Override
05     public String say(String name) {
06         String str = "Hello "+name+", 当前时间是: "+ LocalDateTime.now();
07         return str;
08     }
09 }
```

步骤03 开发服务器。

服务器通过RPC.Builder来创建服务,并通过一个指定的端口对外暴露需要被调用的类。

代码 3.3　RpcServer.java

```
01 package org.hadoop.rpc.server;
02 import org.apache.hadoop.conf.Configuration;
03 import org.apache.hadoop.ipc.RPC;
04 import org.hadoop.rpc.service.HelloImpl;
05 import org.hadoop.rpc.service.IHello;
06 public class RpcServer {
07 public static void main(String[] args) throws Exception {
08     System.setProperty("hadoop.home.dir","D:/program/hadoop-3.2.2");
09     RPC.Server server = new RPC.Builder(new Configuration())
10         .setProtocol(IHello.class)//设置协议即接口
11         .setInstance(new HelloImpl())//设置实例类
12         .setBindAddress("127.0.0.1")//设置服务器地址
13         .setPort(5678)//设置服务端口
14         .build();
15     server.start();//启动RPC服务
16     }
17 }
```

开发完成RpcServer后,执行代码启动RpcServer。启动后,将会占用5678端口,并等待客户端的连接。输出的信息如下所示,即表示启动成功。

```
Starting Socket Reader #1 for port 5678
IPC Server listener on 5678: starting
```

步骤 04 开发客户端。

使用RPC.getProxy获取本地的一个代理，但是接口类必须与服务器的接口类保持一样，如果调用方是另一个项目，则将接口类复制到调用项目中即可。

代码3.4 RpcClient.java

```
01 package org.hadoop.rpc.client;
02 import org.apache.hadoop.conf.Configuration;
03 import org.apache.hadoop.ipc.RPC;
04 import org.hadoop.rpc.service.IHello;
05 import java.net.InetSocketAddress;
06 public class RpcClient {
07     public static void main(String[] args) throws Exception {
08         System.setProperty("hadoop.home.dir","D:/program/hadoop-3.2.2");
09         IHello hello =
10                 RPC.getProxy(IHello.class,//指定协议或接口
11                         1L,//指定客户端ID，任意Long值
12                         new InetSocketAddress("127.0.0.1", 5678),
13                         new Configuration());
14         String str = hello.say("hadoop");
15         System.out.println("返回的数据是: "+str);
16     }
17 }
```

最后运行上面的程序代码，如果可以正常返回数据，则说明RPC远程调用成功。RPC调用的过程就是通过服务器地址和端口实现远程类方法的调用过程。

3.7 小　　结

（1）Hadoop默认的配置在core-default.xml文件中，此文件在hadoop-common.jar包中，其部分配置如下：

```
01 <property>
02     <name>hadoop.common.configuration.version</name>
03     <value>3.0.0</value>
04     <description>version of this configuration file</description>
05 </property>
06 <property>
07     <name>hadoop.tmp.dir</name>
08     <value>/tmp/hadoop-${user.name}</value>
09     <description>A base for other temporary directories.</description>
10 </property>
```

其中，第03行说明了Hadoop的大版本号。第08行，如果没有配置hadoop.tmp.dir，则默认将数据保存到此目录下。

建议不要修改core-default.xml文件中的内容，如果的确需要修改，可以修改${HADOOP_HOME}/etc/hadoop/core-site.xml中的内容。

（2）在hadoop-hdfs.jar中包含有hdfs-default.xml文件，里面保存了默认的HDFS配置，如：

```
01 <property>
02     <name>dfs.replication</name>
03     <value>3</value>
04 </property>
05 <property>
06     <name>dfs.blocksize</name>
07     <value>134217728</value>
08 </property>
```

其中，第03行为默认副本数量；第07行为默认块大小，值为128MB。

（3）Hadoop HDFS进程的具体功能。

（4）RPC远程调用的基本实现。

第 4 章

MapReduce实战

主要内容：

- ❖ MapReduce的执行原理、执行过程。
- ❖ 数据类型及数据格式。
- ❖ Writable接口与序列化机制。
- ❖ WritableComparable实现排序。
- ❖ 默认Mapper和默认Reducer。
- ❖ 倒排索引。
- ❖ MapReduce的源码分析。

4.1 MapReduce 的运算过程

MapReduce为分布式计算模型，分布式计算最早由Google提出。MapReduce将运算的过程分为两个阶段：Map和Reduce。用户只需要实现map和reduce两个函数即可。这两个函数参数的形式都是以Key-Value的形式成对出现，Key为输入或输出的信息，Value的输入或输出的值。

图4-1展示了MapReduce的运算过程。MapReduce将大任务交给多个机器分布式进行计算，然后再进行汇总合并。

图中第1行，Input为输入源，因为输入源是以分布式形式保存到HDFS上的，因此可以同时开启多个Mapper程序，同时读取数据。读取数据时，Mapper将接收两个输入的参数，第一个Key为读取到的文件的行号，Value为读取到的这一行的数据。

第2~3行，Mapper在处理完成以后，也将输出Key-Value对形式的数据。

第4行，是Reduce接收Mapper输入的数据，在接收数据之前先进行排序操作，这个排序操作我们一般称为shuffle。注意，Reduce接收到的Value值是一个数组，多个重复Key的Value会在Reduce程序执行时合并成数组。

第5行，是Reduce的输出，也是以Key-Value对象的形式对外输出的文件或其他存储格式。

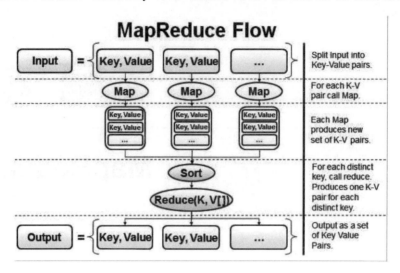

图4-1　MapReduce的运算过程（图片来自于Hadoop官网）

下面以WordCount为例，再为读者讲解一下MapReduce的过程，如图4-2所示。

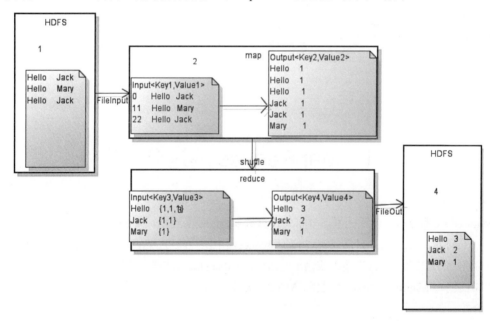

图4-2　MapReduce的过程

标注为1的部分为Hadoop HDFS文件系统中的文件，即被处理的数据首先应该保存到HDFS文件系统中。

标注为2的部分将接收FileInputFormat的输入数据，在处理WordCount示例时，接收到的数据Key1为LongWritable类型，即为字节的偏移量。比如，2中第一行输入为0，其中0为字节0下标的开始，第二行为11，则11为文本中第二行字节的偏移量，以此类推。而Value1则为Text，即文本类型，其中第一行Hello Jack为读取的第一行的数据，以此类推。然后，我们将开发代码对Value1的数据进行处理，以空格或者\t作为分隔，将Hello、Jack等分别输出。此时每一次输出算是一个字符，所以在map中的输出格式是：Key2为Text类型，Value2则为LongWritable类型。

标注为3的部分接收map的输出，所以Key3和Value3的类型应该与Key2和Value2的类型一致。现在我们只需要将Value中的值相加，就可以得到Hello出现的次数。然后直接输出给Key4和Value4，因此Key4、Value4的类型也分别是Text和LongWritable。

最终，数据将保存到HDFS上，即Key4和Value4的数据。如果你已经理解了MapReduce的运算过程，就可以快速开发出WordCount的代码了。

注意：LongWritable和Text为Hadoop中的序列化类型，可以简单地理解为Java中的Long和String。

4.2　WordCount 示例

为了让读者快速掌握MapReduce，本节再次为大家讲解和演示WordCount示例程序，并以本地运行和服务器运行的方式分别部署，让读者快速掌握MapReduce的开发、运行和部署。

之前讲过，MapReduce程序可以运行在本地，也可以打包后运行在Hadoop集群上。之前已经开发过运行在本地的MapReduce程序，这里我们使用打包的方式将程序打包后放到Hadoop集群上运行。

步骤01 创建Java项目添加依赖。

创建Java项目，并添加以下依赖。注意，本次添加的依赖为hadoop-minicluster，且设置scope的值为provided（provided的意思是项目打包时，不会被打包到依赖的jar包中）。

```xml
<dependency>
    <groupId>org.apache.hadoop</groupId>
    <artifactId>hadoop-minicluster</artifactId>
    <scope>provided</scope>
</dependency>
```

步骤02 开发WordCount完整代码。

代码4.1　WordCount.java

```
01 package org.hadoop;
02 import org.apache.hadoop.conf.Configuration;
03 import org.apache.hadoop.conf.Configured;
04 import org.apache.hadoop.fs.FileSystem;
05 import org.apache.hadoop.fs.Path;
06 import org.apache.hadoop.io.LongWritable;
07 import org.apache.hadoop.io.Text;
08 import org.apache.hadoop.mapreduce.Job;
09 import org.apache.hadoop.mapreduce.Mapper;
10 import org.apache.hadoop.mapreduce.Reducer;
11 import org.apache.hadoop.mapreduce.lib.input.FileInputFormat;
12 import org.apache.hadoop.mapreduce.lib.output.FileOutputFormat;
13 import org.apache.hadoop.util.Tool;
14 import org.apache.hadoop.util.ToolRunner;
```

```
15  import java.io.IOException;
16  public class WordCount extends Configured implements Tool {
17      public static void main(String[] args) throws Exception {
18          int result = ToolRunner.run(new WordCount(), args);
19          System.exit(result);
20      }
21      private static String server = "hdfs://server201:8020";
22      public int run(String[] args) throws Exception {
23          if (args.length != 2) {
24              System.err.println("usage: " + this.getClass().getSimpleName() + " <inPath> <outPath>");
25              ToolRunner.printGenericCommandUsage(System.out);
26              return -1;
27          }
28          Configuration config = getConf();
29          config.set("fs.defaultFS", server);
30          //指定resourcemanger的地址
31          config.set("yarn.resourcemanager.hostname", "server201");
32          config.set("dfs.replication", "1");
33          config.set("dfs.permissions.enabled", "false");
34          FileSystem fs = FileSystem.get(config);
35          Path dest = new Path(server + args[1]);
36          if (fs.exists(dest)) {
37              fs.delete(dest, true);
38          }
39          Job job = Job.getInstance(config,"WordCount");
40          job.setJarByClass(getClass());
41          job.setMapperClass(WordCountMapper.class);
42          job.setReducerClass(WordCountReducer.class);
43          job.setOutputKeyClass(Text.class);
44          job.setOutputValueClass(LongWritable.class);
45          FileInputFormat.addInputPath(job, new Path(server + args[0]));
46          FileOutputFormat.setOutputPath(job, dest);
47          boolean boo = job.waitForCompletion(true);
48          return boo ? 0 : 1;
49      }
50      public static class WordCountMapper extends Mapper<LongWritable, Text, Text, LongWritable> {
51          private LongWritable count = new LongWritable(1);
52          private Text text = new Text();
53          @Override
54          public void map(LongWritable key, Text value, Context context) throws IOException, InterruptedException {
55              String str = value.toString();
56              String[] strs = str.split("\\s+");
57              for (String s : strs) {
58                  text.set(s);
59                  context.write(text, count);
60              }
61          }
62      }
63      public static class WordCountReducer extends Reducer<Text, LongWritable, Text, LongWritable> {
64          @Override
65          public void reduce(Text key, Iterable<LongWritable> values, Context context) throws IOException, InterruptedException {
```

```
66              long sum = 0;
67              for (LongWritable w : values) {
68                  sum += w.get();
69              }
70              context.write(key, new LongWritable(sum));
71          }
72      }
73  }
```

上例代码中,由于我们声明了完整的地址,因此可以在本地运行测试。在本地运行测试需要输入两个参数,在IDEA的菜单上选择run→Edit Configurations,并在配置窗口的Program Arguments位置输入读取文件的地址和输出结果的目录,如图4-3所示。

图4-3　输出结果

在本地环境下直接运行项目,并查看HDFS上的结果目录,可以看到WordCount程序已经将结果输出到指定的目录。

```
[hadoop@server201 ~]$ hdfs dfs -ls /out001
Found 2 items
-rw-r--r--   1 wangj supergroup          0 2021-03-13 22:14 /out001/_SUCCESS
-rw-r--r--   1 wangj supergroup        520 2021-03-13 22:14 /out001/part-r-00000
```

步骤 03 使用Maven打包程序。

在IDEA窗口右侧的Maven视图中,单击package并运行,可以得到一个jar包,如图4-4所示。

jar包可以在target目录下找到,将jar包上传到Linux系统,并执行yarn jar命令,如下所示:

```
$ yarn jar chapter04-1.0.jar org.hadoop.WordCount
/test/a.txt /out002
```

查看执行结果,即为单词统计的结果。这个结果根据处理的文件不同,显示的内容会有所不同。

图4-4　单击package并运行

```
[root@server201 ~]# hdfs dfs -cat /out002/* | head
->      4
0       2
1       4
```

至此,我们已经学会如何在本地及打包到服务器上运行MapReduce程序了。在后面的章节中我们将详解MapReduce的更多细节。

注意:在本地运行时,有可能会出现以下错误:

```
Exception in thread "main"
java.lang.UnsatisfiedLinkError: org.apache.hadoop.io.nativeio.
NativeIO$Windows.access0(Ljava/lang/String;I)Z
```

解决方案是：将hadoop.dll文件放到windows/system32目录下即可。

可以将Mapper和Reducer开发成内部类，但这两个内部类必须使用public static修饰符。

上例程序，在IDEA中执行package打包，将得到一个没有任何依赖，只有WordCount代码的jar文件，直接就可以发布到Linux系统的Hadoop集群上执行。因为在Linux上已经存在Hadoop的所有依赖包，所以不需要再将Hadoop的所有依赖都打包到jar文件中去。

4.3 自定义Writable

在Hadoop中，LongWritable、Text都是被序列化的类，它们都是org.apache.hadoop.io.Writable的子类，也只有被序列化的类，才可在Mapper和Reducer之间传递。在实际开发中，为了适应不同业务的要求，有时必须自己开发Writable类的子类，以实现Hadoop中的个性化开发。以下是JDK的序列化与Hadoop序列化的比较。

- JDK的序列化接口为java.io.Serializable，用于将对象转换成字节流输出，即为序列化，再将字节流转换成对象，即为反序列化。
- Hadoop的序列化接口为org.apache.hadoop.io.Writable。它的特点是紧凑（高效的使用存储空间）、快速（读写开销小）、可扩展（可以透明地读取数据）、互操作（支持多语言）。

Writable接口的两个主要方法：write(DataOutput)将成员变量按顺序写出，readFields(DataInput)顺序读取成员变量的值。以下是Writable的源代码。

```
package org.apache.hadoop.io;
public interface Writable {
    void write(DataOutput out) throws IOException;
    void readFields(DataInput in) throws IOException;
}
```

Writable接口的基本实现，就是在write/readFields方法中顺序写出和读取成员变量的值。以下代码是一个实现了Writable接口的具体类，此类省略了setters和getters方法，请注意write和readFields中的代码必须按顺序读写数据。

代码4.2 WritableDemo.java

```
01 package org.hadoop;
02 import org.apache.hadoop.io.Writable;
03 import java.io.DataInput;
04 import java.io.DataOutput;
05 import java.io.IOException;
06 public class WritableDemo implements Writable {
07     private String name;
08     private Integer age;
09     @Override
```

```
10    public void write(DataOutput out) throws IOException {
11        out.writeUTF(name);
12        out.writeInt(age);
13    }
14    @Override
15    public void readFields(DataInput in) throws IOException {
16        name = in.readUTF();
17        age = in.readInt();
18    }
19 }
```

在自定义的类实现接口Writable以后,就可以将这个类作为Key或是Value放到Mapper或是Reducer中当作参数。

现在我们来自己定义一个序列化类,用于统计文本文件中每一行字符的数量。根据这个要求,我们需要定义Mapper的输出类型为自定义的Writable,而Mapper的输出则为Reduce的输入。

步骤01 创建目标读取文件。

首先,先定义一个文本文件,并输入若干行内容。如创建一个a.txt文件在D:/目录下,内容为:

```
Hello This Is First Line .
This Example show how
to implements Writable
```

步骤02 创建Writable类的子类。

由于我们的工作是要读取这一行的数据和这一行有多少字符,所以应该定义两个成员变量,如代码4.3所示。注意,在代码4.3中,需要实现的是Writable的子类WritableComparable<T>,因为我们要将此类作为输出的Key值,则输出的Key必须实现排序的功能,此类WritableComparable的comparaTo方法可以实现排序。

代码 4.3   LineCharCountWritable.java

```
01 package org.hadoop.writable;
02 import org.apache.hadoop.io.WritableComparable;
03 import java.io.DataInput;
04 import java.io.DataOutput;
05 import java.io.IOException;
06 public class LineCharCountWritable implements
07             WritableComparable<LineCharCountWritable> {
08    private String line;
09    private Integer count;
10    @Override
11    public void write(DataOutput out) throws IOException {
12        out.writeUTF(line);
13        out.writeInt(count);
14    }
15    @Override
16    public void readFields(DataInput in) throws IOException {
17        line = in.readUTF();
18        count = in.readInt();
19    }
20    @Override
21    public int compareTo(LineCharCountWritable o) {
```

```
22             return this.count - o.getCount();
23         }
24 }
```

步骤 03 在MapReduce中使用自定义Writable。

在项目中使用自定义的Writable子类,见代码4.4。

代码 4.4　LineCharCountMR.java

```
01 package org.hadoop.writable;
02 import org.apache.commons.lang3.StringUtils;
03 import org.apache.hadoop.conf.Configuration;
04 import org.apache.hadoop.conf.Configured;
05 import org.apache.hadoop.fs.FileSystem;
06 import org.apache.hadoop.fs.Path;
07 import org.apache.hadoop.io.LongWritable;
08 import org.apache.hadoop.io.NullWritable;
09 import org.apache.hadoop.io.Text;
10 import org.apache.hadoop.mapreduce.Job;
11 import org.apache.hadoop.mapreduce.Mapper;
12 import org.apache.hadoop.mapreduce.Reducer;
13 import org.apache.hadoop.mapreduce.lib.input.FileInputFormat;
14 import org.apache.hadoop.mapreduce.lib.output.FileOutputFormat;
15 import org.apache.hadoop.util.Tool;
16 import org.apache.hadoop.util.ToolRunner;
17 import java.io.IOException;
18 public class LineCharCountMR extends Configured implements Tool {
19     public static void main(String[] args) throws Exception {
20         int run = ToolRunner.run(new LineCharCountMR(), args);
21         System.exit(run);
22     }
23     @Override
24     public int run(String[] args) throws Exception {
25         if (args.length < 2) {
26             System.out.println("参数错误,使用方法: LineCharCountMR <Input> <Output>");
27             ToolRunner.printGenericCommandUsage(System.out);
28             return 1;
29         }
30         Configuration config = getConf();
31         FileSystem fs = FileSystem.get(config);
32         Path dest = new Path(args[1]);
33         if (fs.exists(dest)) {
34             fs.delete(dest, true);
35         }
36         Job job = Job.getInstance(config, "LineChar");
37         job.setJarByClass(getClass());
38         job.setMapperClass(LineMapper.class);
39         job.setMapOutputKeyClass(LineCharCountWritable.class);
40         job.setMapOutputValueClass(NullWritable.class);
41         job.setReducerClass(LineReducer.class);
42         job.setOutputKeyClass(Text.class);
43         job.setOutputValueClass(NullWritable.class);
44         FileInputFormat.addInputPath(job, new Path(args[0]));
45         FileOutputFormat.setOutputPath(job, dest);
```

```
46          boolean b = job.waitForCompletion(true);
47          return b ? 0 : 1;
48     }
49     //注意最后一个参数为NullWritable，可以理解为Null
50     public static class LineMapper extends Mapper<LongWritable, Text,
    LineCharCountWritable, NullWritable> {
51         private LineCharCountWritable countWritable = new LineCharCountWritable();
52         @Override
53         protected void map(LongWritable key, Text value, Context context) throws
    IOException, InterruptedException {
54             String line = value.toString();
55             if (StringUtils.isBlank(line)) {
56                 return;
57             }
58             Integer charCount = line.length();
59             countWritable.setLine(line);
60             countWritable.setCount(charCount);
61             context.write(countWritable, NullWritable.get());
62         }
63     }
64     public static class LineReducer extends Reducer<LineCharCountWritable,
    NullWritable, Text, NullWritable> {
65         private Text text = new Text();
66         @Override
67         protected void reduce(LineCharCountWritable key, Iterable<NullWritable>
    values, Context context) throws IOException, InterruptedException {
68             text.set(key.getLine() + "\t" + key.getCount());
69             context.write(text, NullWritable.get());
70         }
71     }
72 }
```

步骤04 运行项目。

现在运行项目，在IDEA的菜单上，单击run→Edit Configuratin，打开配置窗口添加两个参数，如图4-5所示。由于我们是在IDEA中直接运行，即运行在本地，所以传递本地的目录即可，如果运行在Hadoop集群中，请传递HDFS上的文件和目录。

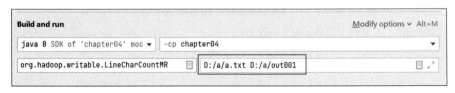

图4-5 添加参数

运行完成后查看输出的结果，如下：

```
This Example show how    21
to implements Writable    22
Hello This Is First Line .    26
```

通过上面的结果可以看出，已经按每行字符数量进行了排序，而这是comparaTo的功能。这个示例演示了如何自定义Writable实现序列化功能。

4.4 Partitioner 分区编程

Partitioner分区编程的主要功能是将不同的分类输出到不同的文件中去。这样在查询数据时，可以根据某个规定的类型查询相关的数据。

使用Partitioner分区，必须给job设置以下两个参数：

（1）设置Reducer的数量，默认为1：

```
job.setNumReduceTasks(2);
```

（2）设置Partitioner类：

```
job.setPartitionerClass(Xxxx.class);
```

Map的结果会通过Partitioner分发到Reducer上，Reducer做完Reduce操作后，通过OutputFormat输出结果，如图4-6所示。

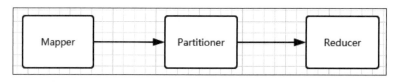

图4-6 输出结果

Mapper最终处理的键值对<Key, Value>需要送到Reducer去合并，当合并时，如果有相同Key的键值对，则会送到同一个Reducer中。任何Key到Reducer的分配过程是由Partitioner决定的，它只有如下所示的一个方法：

```
getPartition(Text key, Text value, int numPartitions)
```

输入参数是Map的结果对<Key, Value>和Reducer的数目，输出则是分配的Reducer（整数编号）的数量。就是指定Mapper输出的键值对到哪一个Reducer上去。系统默认的Partitioner是HashPartitioner，它以Key的Hash值对Reducer的数目取模，得到对应的Reducer。这样可以保证如果有相同的Key值，肯定会被分配到同一个Reducer上。如果有N个Reducer，编号就为0,1,2,3…(N–1)。

在执行job之前，如果设置Reducer的个数为：

```
job.setNumReduceTasks(2);
```

则默认会根据Key的Hash值将数据分别输出到2个文件中，默认使用HashPartitioner类的源码如下：

```
@InterfaceAudience.Public
@InterfaceStability.Stable
public class HashPartitioner<K2, V2> implements Partitioner<K2, V2> {
    public void configure(JobConf job) {}
    /** Use {@link Object#hashCode()} to partition. */
    public int getPartition(K2 key, V2 value,
                     int numReduceTasks) {
        return (key.hashCode() & Integer.MAX_VALUE) % numReduceTasks;
    }
}
```

通过上面的代码算法可知：key.hashCode & Integer.MAX_VALUE是将数据转成正数，然后与设置的Reduce的个数取模。

现在先来做一个测试，在代码4.4中，我们先修改Reducer的个数，给job添加Reducer个数的代码如下所示，其中"..."为省略掉一些无关代码。

```
...
FileOutputFormat.setOutputPath(job, dest);
//设置Reducer的个数
job.setNumReduceTasks(2);
boolean b = job.waitForCompletion(true);
...
```

然后再重新执行代码4.4，查看输出结果，可以看到输出的文件变成两个，且两个文件中的内容不相同。

```
part-r-00000
part-r-00001
```

现在我们可以自己开发Partitioner来修改默认使用hashCode这个规则，如我们使用字符个数来决定输出，将字符个数大于25的输出到一个文件中，小于或等于25个字符的输出到另一个文件中。现在开发Partitioner类的继承类如代码4.5所示。在代码中，第7行开始的语句块，通过判断每一行的字符个数，将数据保存到不同的文件中，通过返回int的值区分保存的文件。

代码 4.5　LineCharCountPartitioner.java

```
01 package org.hadoop.writable;
02 import org.apache.hadoop.io.NullWritable;
03 import org.apache.hadoop.mapreduce.Partitioner;
04 public class LineCharCountPartitioner extends Partitioner<LineCharCountWritable,
   NullWritable> {
05     @Override
06     public int getPartition(LineCharCountWritable key, NullWritable value, int i) {
07         if(key.getCount()>25){
08             return 1;
09         }else{
10             return 0;
11         }
12     }
13 }
```

然后将自定义的Partitioner设置到job中去即可：

```
job.setNumReduceTasks(2);
job.setPartitionerClass(LineCharCountPartitioner.class);
```

运行并查看结果，可以发现字符个数小于25的数据已经输入到part-r-00000文件中，字符个数大于25的数据输出到part-r-00001文件中了。

4.5 自定义排序

在Hadoop开发中，通过Mapper输出给Reducer的数据可以根据Key参数值进行排序，即Mapper<Key1_input,Value1_input,Key2_out,Value2_out>，其中默认根据Key2_out进行排序。

因此，如果希望输出的数据进行排序，可以通过以下方式来实现：

（1）开发JavaBean实现接口WritableComparable，此类是Writable的子类。
（2）将JavaBean作为Mapper的Key2输出给Reducer。
（3）实现JavaBean的comparaTo方法来排序。

在开发时，可以开发多个MapReduce处理过程，形成一个处理链，后面的MapReduce处理前面MapReduce输出的结果，直到最后获得所需要的数据格式。

以下开发一个自定义排序的Wriable类，为大家演示自定义排序的开发过程。

1. 实现WritableComparable接口

之前WordCount输出的数据默认是升序排序的。现在我们可以自定义一个排序规则，让它按倒序排列。

代码4.6　MyText.java

```
01 package org.hadoop.sort;
02 import org.apache.hadoop.io.WritableComparable;
03 import java.io.DataInput;
04 import java.io.DataOutput;
05 import java.io.IOException;
06 public class MyText implements WritableComparable<MyText> {
07     private String text;
08     @Override
09     public int compareTo(MyText o) {
10         return o.text.compareTo(this.text);
11     }
12     @Override
13     public void write(DataOutput out) throws IOException {
14         out.writeUTF(text);
15     }
16     @Override
17     public void readFields(DataInput in) throws IOException {
18         text = in.readUTF();
19     }
20     public String getText() {
21         return text;
22     }
23     public void setText(String text) {
24         this.text = text;
25     }
26 }
```

然后将MyText作为Mapper的输出Key值。首先设置job输出的格式：

```
job.setMapOutputKeyClass(MyText.class);
job.setMapOutputValueClass(LongWritable.class);
```

设置为Mapper的输出：

```
public static class WordCountMapper extends Mapper<LongWritable, Text, MyText, LongWritable> { ... }
```

设置Reducer接收MyText作为输入：

```
public static class WordCountReducer extends Reducer<MyText, LongWritable, Text, LongWritable> { .. }
```

执行代码，查看输出结果，可以看到数据已经是倒序排列了：

```
wangjian        1
value.toString();  1
throws       4
text.set(s);    1
```

2. 实现WritableComparator接口

如果不希望修改Mapper输出的Key值部分，则可以定义一个Comparator，如代码4.7所示。

代码 4.7　MyComparator.java

```
01 package org.hadoop.sort;
02 import org.apache.hadoop.io.Text;
03 import org.apache.hadoop.io.WritableComparable;
04 import org.apache.hadoop.io.WritableComparator;
05 public class MyComparator extends WritableComparator {
06     public MyComparator() {
07         //注意构造部分，第二个true必须传递，否则会抛出异常
08         super(Text.class,true);
09     }
10     @Override
11     public int compare(WritableComparable a, WritableComparable b) {
12         int c = b.compareTo(a);
13         return c;
14     }
15 }
```

然后在job中添加一个排序对象即可：

```
job.setSortComparatorClass(MyComparator.class);
```

最后，执行代码并查看结果，就可以看到结果会按指定的顺序显示。

4.6　Combiner 编程

Combiner指在Mapper以后先对数据进行一个计算，然后再将数据发送到Reducer。开发Combiner

就是开发一个继承Reducer的类来实现Reducer中相同的功能。Combiner是可插拔的组件，但仅限于Combiner的业务与Reducer相同的业务。在开发的代码中，可以通过调用job.setCombinerClass(..)来添加Combiner类。

以下示例用来说明没有Combiner和有Combiner两种不同的执行过程：

源数据：

```
Hello    Jack
Hello    Mary
Hello    Jack
Hello    Jack
```

没有Combiner的情况：

```
Mapper              Shuffle                 Reducer
<Hello,1>           <Hello,[1,1,1,1]>       <Hello,4>
<Jack,1>            <Jack,[1,1,1]>          <Jack,3>
<Hello,1>            <Mary,[1]>             <Mary,1>
<Mary,1>
<Hello,1>
<Jack,1>
<Hello,1>
<Jack,1>
```

有Combiner的情况：

```
Mapper          Combiner        Shuffle             Reducer
<Hello,1>       <Hello,4>       <Hello,[4]>         <Hello,4>
<Jack,1>        <Jack,3>        <Jack,[3]>          <Jack,3>
<Hello,1>       <Mary,1>         <Mary,[1]>         <Mary,1>
<Mary,1>
<Hello,1>
<Jack,1>
<Hello,1>
<Jack,1>
```

通过上面的分析可以看到，在Combiner中已经对数据进行了一次合并计算操作，这将减少在最后的Reducer中遍历处理的次数。

现在添加这个Combiner，在启动之前调用setCombinerClass(..)即可。

```
// 设置一个Combiner
job.setCombinerClass(PhoneDataReducer.class);
// 开始任务
job.waitForCompletion(true);
```

由于Combiner类就是Reducer类的子类，所以此处就不再赘述。读者可以直接将Reducer类作为Combiner设置到job中去即可。

4.7　默认Mapper和默认Reducer

如果没有给job设置Mapper和Reducer，则job将会使用默认的Mapper和Reducer。默认的Mapper和

Reducer什么都不会做,只是进行文件的快速复制。如代码4.8所示,此代码并没有设置Mapper和Reducer类,将会使用默认的Mapper和Reducer类。

代码4.8　DefaultMR.java

```
01 package org.hadoop.dft;
02 import org.apache.hadoop.conf.Configuration;
03 import org.apache.hadoop.conf.Configured;
04 import org.apache.hadoop.fs.FileSystem;
05 import org.apache.hadoop.fs.Path;
06 import org.apache.hadoop.mapreduce.Job;
07 import org.apache.hadoop.mapreduce.lib.input.FileInputFormat;
08 import org.apache.hadoop.mapreduce.lib.output.FileOutputFormat;
09 import org.apache.hadoop.util.Tool;
10 import org.apache.hadoop.util.ToolRunner;
11 public class DefaultMR extends Configured implements Tool {
12     @Override
13     public int run(String[] args) throws Exception {
14         if(args.length!=2){
15             System.out.println("usage : <in> <out>");
16             return -1;
17         }
18         Configuration conf = getConf();
19         Job job = Job.getInstance(conf,"DefaultMR");
20         job.setJarByClass(this.getClass());
21         FileSystem fs = FileSystem.get(conf);
22         if(fs.exists(new Path(args[1]))){
23             fs.delete(new Path(args[1]),true);
24         }
25         FileInputFormat.setInputPaths(job,new Path(args[0]));
26         FileOutputFormat.setOutputPath(job,new Path(args[1]));
27         int code = job.waitForCompletion(true)?0:1;
28         return code;
29     }
30     public static void main(String[] args) throws Exception {
31         int code = ToolRunner.run(new DefaultMR(),args);
32         System.exit(code);
33     }
34 }
```

4.8　倒 排 索 引

倒排索引用于统计并记录某个单词在一个文件中出现的次数及位置。我们可以实现一个简单的算法来统计单词在一个文件中出现的次数。假如存在以下两个文件:

(1) a.txt文件的内容为:

```
Hello Jack
Hello Jack
```

（2）b.txt文件的内容为：

```
Hello Mary
```

则统计完成以后的结果为：

```
单词        文件        出现次数    文件        出现次数    总出现次数
Hello      a.txt, 2               b.txt, 1               3
Jack       a.txt, 2                                      2
Mary                              b.txt, 1               1
```

处理的思路可以是先根据Word+文件名做一次统计，结果为：

```
单词        文件        出现次数
Hello      a.txt       2
Hello      b.txt       1
Jack       a.txt       2
Mary       b.txt       1
```

然后再对上面的结果进行处理，以Word为Key，以文件名、次数为Value进行再处理，并最终输出要求的结果。

第一个MapReduce程序，用于将两个文件中的数据先按单词+\t+文件+\t+出现次数进行统计。

代码 4.9　InverseMR1.java

```
01  package org.hadoop.invise;
02  import org.apache.hadoop.conf.Configuration;
03  import org.apache.hadoop.conf.Configured;
04  import org.apache.hadoop.fs.FileSystem;
05  import org.apache.hadoop.fs.Path;
06  import org.apache.hadoop.io.LongWritable;
07  import org.apache.hadoop.io.NullWritable;
08  import org.apache.hadoop.io.Text;
09  import org.apache.hadoop.mapreduce.InputSplit;
10  import org.apache.hadoop.mapreduce.Job;
11  import org.apache.hadoop.mapreduce.Mapper;
12  import org.apache.hadoop.mapreduce.Reducer;
13  import org.apache.hadoop.mapreduce.lib.input.FileInputFormat;
14  import org.apache.hadoop.mapreduce.lib.input.FileSplit;
15  import org.apache.hadoop.mapreduce.lib.input.TextInputFormat;
16  import org.apache.hadoop.mapreduce.lib.output.FileOutputFormat;
17  import org.apache.hadoop.mapreduce.lib.output.TextOutputFormat;
18  import org.apache.hadoop.util.Tool;
19  import org.apache.hadoop.util.ToolRunner;
20  import java.io.IOException;
21  public class InverseMR1 extends Configured implements Tool {
22      @Override
23      public int run(String[] args) throws Exception {
24          Configuration conf = getConf();
25          FileSystem fs = FileSystem.get(conf);
26          Path dest = new Path("D:/a/out002");
27          if (fs.exists(dest)) {
28              fs.delete(dest, true);
29          }
30          Job job = Job.getInstance(conf, "InverseIndex");
31          job.setJarByClass(getClass());
32          job.setMapperClass(IIMapper.class);
33          job.setMapOutputKeyClass(Text.class);
```

```
34          job.setMapOutputValueClass(LongWritable.class);
35          job.setReducerClass(IIReducer.class);
36          job.setOutputKeyClass(Text.class);
37          job.setOutputValueClass(NullWritable.class);
38          job.setInputFormatClass(TextInputFormat.class);
39          job.setOutputFormatClass(TextOutputFormat.class);
40          FileInputFormat.setInputPaths(job, new Path("D:/a/in"));
41          FileOutputFormat.setOutputPath(job, dest);
42          int code = job.waitForCompletion(true) ? 0 : 1;
43          return code;
44      }
45      public static class IIMapper extends Mapper<LongWritable, Text, Text, LongWritable> {
46          private String fileName = "";
47          private Text key = new Text();
48          private LongWritable value = new LongWritable(0L);
49          @Override
50          public void map(LongWritable key, Text value, Context context) throws IOException, InterruptedException {
51              String[] strs = value.toString().split("\\s+");
52              for (String str : strs) {
53                  this.key.set(str + "\t" + fileName);
54                  this.value.set(1L);
55                  context.write(this.key, this.value);
56              }
57          }
58          @Override
59          protected void setup(Context context) throws IOException, InterruptedException {
60              InputSplit split = context.getInputSplit();
61              if (split instanceof FileSplit) {
62                  FileSplit fileSplit = (FileSplit) split;
63                  fileName = fileSplit.getPath().getName();
64              }
65          }
66      }
67      public static class IIReducer extends Reducer<Text, LongWritable, Text, NullWritable> {
68          @Override
69          public void reduce(Text key, Iterable<LongWritable> values, Context context) throws IOException, InterruptedException {
70              long sum = 0L;
71              for (LongWritable l : values) {
72                  sum += l.get();
73              }
74              key.set(key.toString() + "\t" + sum);
75              context.write(key, NullWritable.get());
76          }
77      }
78      public static void main(String[] args) throws Exception {
79          int code = ToolRunner.run(new InverseMR1(), args);
80          System.exit(code);
81      }
82  }
```

统计后的结果如下:

```
Hello   a.txt   2
Hello   b.txt   1
Jack    a.txt   2
Mary    b.txt   1
```

第二个MapReduce程序，用于将上面的结果再根据单词进行统计，如代码4.10所示。

代码4.10　InverseMR2.java

```java
01  package org.hadoop.inverse;
02  import org.apache.hadoop.conf.Configuration;
03  import org.apache.hadoop.conf.Configured;
04  import org.apache.hadoop.fs.FileSystem;
05  import org.apache.hadoop.fs.Path;
06  import org.apache.hadoop.io.LongWritable;
07  import org.apache.hadoop.io.Text;
08  import org.apache.hadoop.mapreduce.Job;
09  import org.apache.hadoop.mapreduce.Mapper;
10  import org.apache.hadoop.mapreduce.Reducer;
11  import org.apache.hadoop.mapreduce.lib.input.FileInputFormat;
12  import org.apache.hadoop.mapreduce.lib.input.TextInputFormat;
13  import org.apache.hadoop.mapreduce.lib.output.FileOutputFormat;
14  import org.apache.hadoop.mapreduce.lib.output.TextOutputFormat;
15  import org.apache.hadoop.util.Tool;
16  import org.apache.hadoop.util.ToolRunner;
17  import java.io.IOException;
18  public class InverseMR2 extends Configured implements Tool {
19      @Override
20      public int run(String[] args) throws Exception {
21          Configuration conf = getConf();
22          FileSystem fs = FileSystem.get(conf);
23          Path dest = new Path("D:/a/out003");
24          if (fs.exists(dest)) {
25              fs.delete(dest, true);
26          }
27          Job job = Job.getInstance(conf, "InverseIndex2");
28          job.setJarByClass(getClass());
29          job.setMapperClass(IIMapper2.class);
30          job.setMapOutputKeyClass(Text.class);
31          job.setMapOutputValueClass(Text.class);
32          job.setReducerClass(IIReducer2.class);
33          job.setOutputKeyClass(Text.class);
34          job.setOutputValueClass(LongWritable.class);
35          job.setInputFormatClass(TextInputFormat.class);
36          job.setOutputFormatClass(TextOutputFormat.class);
37          FileInputFormat.setInputPaths(job, new Path("D:/a/out002"));
38          FileOutputFormat.setOutputPath(job, dest);
39          int code = job.waitForCompletion(true) ? 0 : 1;
40          return code;
41      }
42      public static class IIMapper2 extends Mapper<LongWritable, Text, Text, Text> {
43          private Text key = new Text();
44          private Text value = new Text();
45          @Override
46          public void map(LongWritable key, Text value, Context context) throws
    IOException, InterruptedException {
```

```
47              String[] strs = value.toString().split("\\s+");
48              this.key.set(strs[0]);//Hello
49              this.value.set(strs[1] + "\t" + strs[2]);//a.txt,1
50              context.write(this.key, this.value);
51          }
52      }
53      public static class IIReducer2 extends Reducer<Text, Text, Text, LongWritable> {
54          private LongWritable sum = new LongWritable(0L);
55          @Override
56          public void reduce(Text key, Iterable<Text> values, Context context) throws
    IOException, InterruptedException {
57              this.sum.set(0L);
58              String str = "";
59              for (Text t : values) {
60                  String[] strs = t.toString().split("\t");
61                  this.sum.set(this.sum.get() + Long.parseLong(strs[1]));
62                  str += "\t" + t.toString();
63              }
64              key.set(key.toString() + "\t" + str);
65              context.write(key, this.sum);
66          }
67      }
68      public static void main(String[] args) throws Exception {
69          int code = ToolRunner.run(new InverseMR2(), args);
70          System.exit(code);
71      }
72  }
```

执行后的结果如下：

```
Hello       b.txt       1       a.txt       2       3
Jack        a.txt       2       2
Mary        b.txt       1       1
```

4.9 Shuffle

　　Shuffle是MapReduce的核心。Shuffle的本义是洗牌、混洗，指的是把一组有一定规则的数据尽量转换成一组无规则的数据，越随机越好。MapReduce中的Shuffle更像是洗牌的逆过程，把一组无规则的数据尽量转换成一组具有一定规则的数据。

　　为什么MapReduce计算模型需要Shuffle过程？我们都知道MapReduce计算模型一般包括两个重要的阶段：Map是映射，负责数据的过滤分发；Reduce是规约，负责数据的计算归并。Reduce的数据来源于Map，Map的输出即是Reduce的输入，Reduce需要通过Shuffle来获取数据。

　　从Map输出到Reduce输入的整个过程，可以广义地称为Shuffle。Shuffle横跨Map端和Reduce端，在Map端包括spill过程，在Reduce端包括copy和sort过程，如图4-7所示。

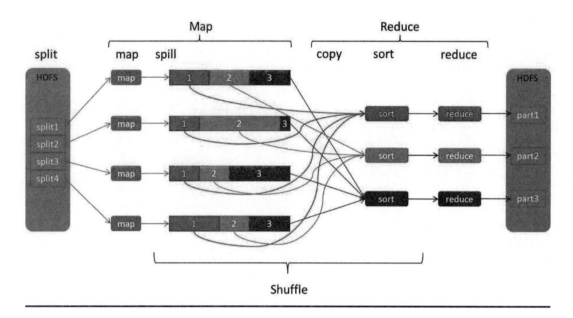

图4-7　Map输出到Reduce输入的过程

1. Spill

Spill过程包括输出、排序、溢写、合并等步骤，如图4-8所示。

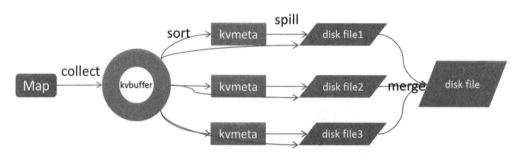

图4-8　Spill过程

Spill的第一个阶段Collect，每个Map任务不断地以<Key, Value="">对的形式把数据输出到在内存中构造的一个环形数据结构中。使用环形数据结构是为了更有效地使用内存空间，在内存中放置尽可能多的数据。

这个数据结构其实就是个字节数组，叫Kvbuffer，名如其义，但是这里面不光放置了<Key, Value="">数据，还放置了一些索引数据，我们给放置索引数据的区域起了一个Kvmeta的别名，在Kvbuffer的一块区域上穿了一个IntBuffer（字节序采用的是平台自身的字节序）的马甲。在Kvbuffer中，<Key, Value="">数据区域和索引数据区域是相邻不重叠的两个区域，用一个分界点来划分两者，分界点不是亘古不变的，而是每次Spill之后都会更新一次。初始的分界点是0，<Key, Value="">数据的存储方向是向上增长，索引数据的存储方向是向下增长。

Kvbuffer的存放指针bufindex是一直向上增长。比如，bufindex初始值为0，一个Int型的Key写完之后，bufindex增长为4；一个Int型的Value写完之后，bufindex增长为8。

索引是对<key, value="">在kvbuffer中的索引，是个四元组，包括：Value的起始位置、Key的起始位置、Partition值、Value的长度，占用四个Int长度。Kvmeta的存放指针Kvindex每次都是向下跳四个"格子"，然后再向上一个格子一个格子地填充四元组的数据。比如Kvindex初始位置是-4，当第一个<key, value="">写完之后,(Kvindex+0)的位置存放value的起始位置、(Kvindex+1)的位置存放key的起始位置、(Kvindex+2)的位置存放Partition的值、(Kvindex+3)的位置存放Value的长度，然后Kvindex跳到-8位置，等第二个<key, value="">和索引写完之后，Kvindex跳到-32位置。

Kvbuffer的大小虽然可以通过参数设置，但是总共就那么大，<key, value="">和索引不断地增加，加着加着，Kvbuffer总有不够用的那一天，这怎么办？把数据从内存刷到磁盘上再接着往内存写数据，把Kvbuffer中的数据刷到磁盘上的过程就叫Spill，多么明了的叫法，内存中的数据满了就自动地Spill到具有更大空间的磁盘上。

关于Spill触发的条件，也就是Kvbuffer用到什么程度开始Spill，还是要讲究一下的。如果把Kvbuffer用得比较满，一点缝都不剩的时候再开始Spill，那Map任务就需要等Spill完成腾出空间之后才能继续写数据；如果Kvbuffer只是满到一定程度，比如80%的时候就开始Spill，那在Spill的同时，Map任务还能继续写数据，如果Spill够快，Map可能都不需要为空闲空间而发愁。两利相衡取其大，一般选择后者。

Spill这个重要的过程是由Spill线程来承担的，Spill线程从Map任务接到"命令"之后就开始正式工作，工作内容叫SortAndSpill，原来不仅仅是Spill，在Spill之前还有个颇具争议性的Sort。

2. Sort

Sort 先把Kvbuffer中的数据按照Partition值和Key两个关键字升序排序，移动的只是索引数据，排序结果是Kvmeta中数据按照Partition为单位聚集在一起，同一Partition内按照Key有序排列。

Spill线程为这次Spill过程创建一个磁盘文件：从所有的本地目录中轮训查找能存储这么大空间的目录，找到之后在其中创建一个类似于"spill12.out"的文件。Spill线程根据排过序的Kvmeta挨个Partition的把<key, value="">数据"吐"到这个文件中，一个Partition对应的数据吐完之后顺序地吐下一个Partition，直到把所有的Partition遍历完。一个Partition在文件中对应的数据也叫段（segment）。

所有的Partition对应的数据都放在这个文件里，虽然是顺序存放的，但是怎么直接知道某个Partition在这个文件中存放的起始位置呢？强大的索引又出场了。有一个三元组记录某个Partition对应的数据在这个文件中的索引：起始位置、原始数据长度、压缩之后的数据长度，一个Partition对应一个三元组。然后把这些索引信息存放在内存中，如果内存中放不下了，后续的索引信息就需要写到磁盘文件中了：从所有的本地目录中轮询查找能存储这么大空间的目录，找到之后在其中创建一个类似于"spill12.out.index"的文件，文件中不光存储了索引数据，还存储了crc32的校验数据（spill12.out.index不一定在磁盘上创建，如果内存（默认1MB空间）中能放得下，就放在内存中，如果内存中放不下，就要写到磁盘）。

每一次Spill过程最少会生成一个out文件，有时还会生成index文件，Spill的次数也烙印在文件名中。索引文件和数据文件的对应关系如图4-9所示。

在Spill线程如火如荼地进行SortAndSpill工作的同时，Map任务不会因此而停歇，而是一如既往地进行着数据输出。Map还是把数据写到Kvbuffer中，那问题就来了：<key,

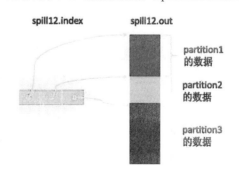

图4-9 索引文件和数据文件的对应关系

value="">只顾着闷头按照bufindex指针向上增长，kvmeta只顾着按照Kvindex向下增长，是保持指针起始位置不变继续跑呢，还是另谋它路？如果保持指针起始位置不变，很快bufindex和Kvindex就碰头了，碰头之后再重新开始或者移动内存都比较麻烦，这个不可取。Map取Kvbuffer中剩余空间的中间位置，用这个位置设置为新的分界点，bufindex指针移动到这个分界点，Kvindex移动到这个分界点的-16位置，然后两者就可以和谐地按照自己既定的轨迹放置数据了，当Spill完成，空间腾出之后，不需要做任何改动继续前进。

Map任务总要把输出的数据写到磁盘上，即使输出数据量很小，在内存中全部能装得下，在最后也会把数据写入磁盘上。

3. Merge

Map任务如果输出数据量很大，可能会进行好几次Spill，out文件和Index文件会产生很多，分布在不同的磁盘上。最后把这些文件进行合并的Merge过程闪亮登场。

Merge过程如何知道产生的Spill文件都在哪了呢？从所有的本地目录上扫描得到产生的Spill文件，然后把路径存储在一个数组里。Merge过程又怎么知道Spill的索引信息呢？没错，也是从所有的本地目录上扫描得到Index文件，然后把索引信息存储在一个列表里。这里，又遇到了一个值得思考的地方。在之前Spill过程的时候为什么不直接把这些信息存储在内存中呢，何必又多了这步扫描的操作？特别是Spill的索引数据，之前当内存超限之后就把数据写到磁盘，现在又要从磁盘把这些数据读出来，还是需要装到更多的内存中。之所以多此一举，是因为此时Kvbuffer这个内存大户已经不再使用，可以回收，有内存空间来装这些数据了。然后为Merge过程创建一个叫file.out的文件和一个叫file.out.Index的文件用来存储最终的输出和索引，一个Partition一个Partition地进行合并输出。对于某个Partition来说，从索引列表中查询这个Partition对应的所有索引信息，每个对应一个段插入到段列表中。也就是这个Partition对应一个段列表，记录所有的Spill文件中对应的这个Partition那段数据的文件名、起始位置、长度等。

然后对这个Partition对应的所有的segment进行合并，目标是合并成一个segment。当这个Partition对应很多个segment时，会分批地进行合并：先从segment列表中把第一批取出来，以Key为关键字放置成最小堆，然后从最小堆中每次取出最小的<Key, Value="">输出到一个临时文件中，这样就把这一批段合并成一个临时的段，把它加回到segment列表中；再从segment列表中把第二批取出来，合并输出到一个临时segment，把其加入到列表中；这样往复执行，直到剩下的段是一批，输出到最终的文件中。

4. Copy

Reduce任务通过HTTP向各个Map任务拖取它所需要的数据。每个节点都会启动一个常驻的HTTP server，其中一项服务就是响应Reduce拖取Map数据。当有MapOutput的HTTP请求过来的时候，HTTP server就读取相应的Map输出文件中对应这个Reduce部分的数据，通过网络流输出给Reduce。

Reduce任务拖取某个Map对应的数据，如果在内存中能放得下这次数据的话，就直接把数据写到内存中。Reduce要向每个Map去拖取数据，在内存中每个Map对应一块数据，当内存中存储的Map数据占用空间达到一定程度的时候，开始启动内存中Merge，把内存中的数据Merge输出到磁盘上的一个文件中。

如果在内存中放不下这个Map的数据的话，直接把Map数据写到磁盘上，在本地目录创建一个文件，从HTTP流中读取数据然后写到磁盘，使用的缓存区大小是64KB。拖一个Map数据过来就会创建一个文件，当文件数量达到一定阈值时，开始启动磁盘文件Merge，把这些文件合并输出到一个文件。

有些Map的数据较小，是可以放在内存中的，有些Map的数据较大，需要放在磁盘上。这样最后Reduce任务拖过来的数据，有些放在内存中，有些放在磁盘上，最后会对这些数据进行全局合并。

5. Merge排序

这里使用的Merge和Map端使用的Merge过程一样。Map的输出数据已经是有序的，Merge进行一次合并排序，所谓Reduce端的Sort过程就是这个合并的过程。一般Reduce是一边复制一边排序，即复制和排序两个阶段是重叠，而不是完全分开的。

Reduce端的Shuffle过程至此结束。

4.10 小　　结

- MapReduce过程被显式地分为两部分：Mapper和Reducer。
- 在MapReduce中被传递的类必须实现Hadoop的序列化接口Writable。
- Mapper中输出的key可以实现排序，此类必须实现接口WritableComparable。
- 多个MapReduce可以分别打成不同的jar包执行。这样后面的MapReduce可以处前面MapReduce输出的数据。
- Partitioner编程用于将不同规则的数据输出到指定编号的Reduce上去。
- Combiner类似于一个前置的Reduce，用于在将数据输出到Reduce之前进行一次数据的合并操作。
- MapReduce的开发过程：

  ① 研究业务需要输出的格式。
  ② 自定义一个类继承 Mapper类，并指定输入输出的格式。
  ③ 重写map方法实现具体的业务逻辑。注意使用context执行输出操作。
  ④ 自定义一个类继承Reduce类，并指定输入输出格式。
  ⑤ 重写Reduce方法。实现自己的业务代码。注意使用context执行输出操作。
  ⑥ 开发一个main方法，通过job对象进行组装。
  ⑦ 打成jar包，指定主类，发送到Linux上，通过yarn jar命令来启动MapReduce。

- MapReduce的执行流程：

  ① Client通过RPC将请求提交给ResourceManager。
  ② ResourceManager在接收到请求以后返回一个jobID给Client。
  ③ Client将jar包上传到HDFS（默认HDFS保存10份），且程序执行完成以后删除jar文件。
  ④ Client通知ResourceManager保存数据的描述信息。
  ⑤ ResourceManager通过公平调度开启任务，将任务放到任务调度队列。
  ⑥ NodeManager通过心跳机制向ResourceManager领取任务。
  ⑦ NodeManager向HDFS领取所需要的jar包，并开始执行任务。

- Partitioner分区编程，就是控制将不同的结果输出到指定的文件中。
- 开启一个Reducer就会拥有一个输出文件。通过设置job.setNumReduceTask，可以设置开启Reducer的个数。此数值必须大于或等于Partitioner分区返回的数量。
- 当启动的Reducer大于Partitioner返回的数量时，将会生成一些空的文件。当Reducer数量小于Partitioner返回的数量时，将直接抛出异常Illege Partition counter error。
- 在设置Reducer数量时，应该考虑Partitioner返回的分区的个数。
- Mapper在输出数据给Reducer时，根据不同的分区编号分发到不同的Reducer上去。每一个Reducer都会有一个编号。

第 5 章

ZooKeeper与高可用集群实战

主要内容：

- ❖ ZooKeeper简介。
- ❖ ZooKeeper快速安装及基本命令。
- ❖ ZooKeeper分布式安装。
- ❖ ZooKeeper的观察节点。
- ❖ Java客户端操作ZooKeeper。
- ❖ 将文件内容保存到ZooKeeper的节点中。
- ❖ 使用ZooKeeper实现Hadoop的高可用（HA）集群。

5.1 ZooKeeper 简介

ZooKeeper是一个分布式的、开放源码的分布式应用程序协调服务，它包含一个简单的原语集，分布式应用程序可以基于它实现同步服务，配置、维护和命名服务等。ZooKeeper是Hadoop的一个子项目。在分布式应用中，由于工程师不能很好地使用锁机制，以及基于消息的协调机制不适合在某些应用中使用，因此需要有一种可靠的、可扩展的、分布式的、可配置的协调机制来统一系统的状态。ZooKeeper的目的就在于此。

目前，在分布式协调技术方面做得比较好的就是Google的Chubby，还有Apache的ZooKeeper，它们都是分布式锁的实现者。有人会问，既然有了Chubby，为什么还要弄一个ZooKeeper，难道Chubby做得不够好吗？不是这样的，主要是Chubby是非开源的，Google自家用。后来雅虎模仿Chubby开发出了ZooKeeper，也实现了类似的分布式锁的功能，并且将ZooKeeper作为一种开源的程序捐献给了Apache，那么这样就可以使用ZooKeeper所提供锁服务。而且，ZooKeeper在分布式领域久经考验，它的可靠性、可用性都是经过理论和实践验证的。因此，我们在构建一些分布式系统的时候，就可以以这类系统为起点来构建我们的系统，这将节省不少成本，而且BUG也将更少。

本章学习ZooKeeper的主要目的就是为了搭建Hadoop的高可用集群。图5-1所示为Google的Chubby和Apache的ZooKeeper示意图。

图5-1　Chubby和ZooKeeper示意图

ZooKeeper中的角色主要有三类，如表5-1所示。

表5-1　ZooKeeper中的角色及说明

角 色		说　明
领导者（Leader）		负责进行投票的发起和决议，更新系统状态
学习者（Learner）	跟随者（Follower）	Follower 用于接收客户请求并向客户返回结果，且在选主过程中参考投票
	观察者（Observer）	Observer 可以接收客户端连接，将写请求转发给 Leader 节点。但 Observer 不参与投票过程，只同步 Leader 的状态。Observer 的目的是为了扩展系统，提高读取速度
客户端（Client）		请求发起方

ZooKeeper的系统模型如图5-2所示。

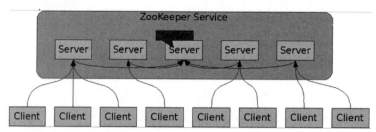

图5-2　ZooKeeper的系统模型

ZooKeeper中有多种记录时间的形式，其中包含以下几个主要属性：Zxid、版本号。

5.1.1　Zxid

致使ZooKeeper节点状态改变的每一个操作，都将使节点接收到一个Zxid格式的时间戳，并且这个时间戳全局有序。也就是说，每个对节点的改变都将产生一个唯一的Zxid。如果Zxid1的值小于Zxid2的值，那么Zxid1所对应的事件发生在Zxid2所对应的事件之前。实际上，ZooKeeper的每个节点维护着三个Zxid值，为别为：cZxid、mZxid和pZxid。

- cZxid：是节点的创建时间所对应的Zxid格式时间戳。
- mZxid：是节点的修改时间所对应的Zxid格式时间戳。
- pZxid：是该节点的子节点列表最后一次被修改时的时间，子节点内容变更不会变更pZxid。

实现中Zxid是一个64位的数字，其高32位是epoch用来标识Leader关系是否改变，每次一个Leader被选出来，它都会有一个新的epoch；其低32位是个递增计数。

5.1.2 版本号

对节点的每一个操作都将导致这个节点的版本号增加。每个节点维护着三个版本号,分别为:

- version:节点数据版本号。
- cversion:子节点版本号。
- aversion:节点所拥有的ACL版本号。

ZooKeeper节点属性。一个节点自身拥有表示其状态的许多重要属性,如表5-2所示。

表5-2 节点自身拥有表示其状态的许多重要属性

属 性	说 明
czxid	节点被创建的 zxid
mzxid	节点被修改的 zxid
ctime	节点被创建的时间
mtime	节点被修改的时间
version	节点被修改的版本号
cversion	节点所拥有的子节点被修改的版本号
aversion	节点的 ACL 被修改的版本号
ephemeralOwner	如果此节点为临时节点,那么它的值为这个节点拥有者的会话 ID,否则值为 0
dataLength	节点数的长度
numChildren	节点拥有的子节点的长度
pzxid	最新修改的 zxid

ZooKeeper服务的操作。在ZooKeeper中有9个基本操作,如表5-3所示。

表5-3 ZooKeeper 中的 9 个基本操作

操 作	说 明
create	创建 znode(父 znode 必须存在)
delete	删除 znode(znode 没有子节点)
exists	判断 znode 是否存在
getACL/setACL	读取或设置 ACL
getChildren	获取 znode 所有子节点的列表
getData/setData	获取/设置 znode 相关数据
sync	客户端与 ZooKeeper 同步

ZooKeeper的配置文件。ZooKeeper的默认配置文件为$ZOOKEEPER_HOME/conf/zoo.cfg。其中各配置项的含义如下:

(1)tickTime:CS通信心跳时间

ZooKeeper 服务器之间或客户端与服务器之间维持心跳的时间间隔,也就是每过tickTime时间就会发送一个心跳。tickTime以毫秒为单位。如tickTime=2000。

（2）initLimit：LF初始通信时限

集群中的Follower服务器（F）与Leader服务器（L）之间初始连接时能容忍的最多心跳数（tickTime的数量）。如initLimit=5。

（3）syncLimit：LF同步通信时限

集群中的Follower服务器与Leader服务器之间请求和应答之间能容忍的最多心跳数（tickTime的数量）。如syncLimit=2。

（4）dataDir：数据文件目录

ZooKeeper保存数据的目录，默认情况下，ZooKeeper将写数据的日志文件也保存在这个目录里。如设置dataDir=/home/zoo/SomeData。

（5）clientPort：客户端连接端口

客户端连接ZooKeeper 服务器的端口，ZooKeeper会监听这个端口，接受客户端的访问请求。如clientPort=2181。

（6）服务器名称与地址

集群信息（服务器编号，服务器地址，LF通信端口，选举端口）。

这个配置项的书写格式比较特殊，规则如下：

```
server.N=YYY:A:B
```

如以下配置示例：

```
server.1=ip:2888:3888
server.2=ip:2888:3888
server.3=ip:2888:3888
```

5.2　单一节点安装 ZooKeeper

首先确认在安装ZooKeeper之前已经安装了JDK，并正确配置了JAVA_HOME和PATH环境变量。单一节点安装只要解压ZooKeeper、配置zoo.cfg文件，并修改dataDir数据保存目录即可。

步骤01 下载ZooKeeper。

ZooKeeper可以通过wget下载。如果没有安装wget，则可以通过yum install -y wget命令安装此软件，也可以通过xftp上传已经下载好的ZooKeeper压缩文件。

下载地址：https://mirrors.bfsu.edu.cn/apache/zookeeper/zookeeper-3.6.2/apache-zookeeper-3.6.2-bin.tar.gz。

步骤02 解压。

解压到指定的目录或当前目录：

```
$ tar -zxvf ~/apache-zookeeper-3.6.2-bin.tar.gz -C /app/
```

步骤 03 修改配置文件设置dataDir目录。

修改示例的配置文件,进入zookeeper目录,将zoo_sample.cfg修改成zoo.cfg文件。

```
$ cp zoo_sample.cfg zoo.cfg
vim zoo.cfg
dataDir=/app/datas/zk
```

步骤 04 启动ZooKeeper。

ZooKeeper默认使用2181端口。启动zkServer。

```
[hadoop@server101 app]$ /app/zookeeper/bin/zkServer.sh start
ZooKeeper JMX enabled by default
Using config: /app/zookeeper/bin/../conf/zoo.cfg
Starting zookeeper ... STARTED
```

查看zk的进程,其中的QuorumPeerMain就是zkServer的进程。

```
$ jps
2490 QuorumPeerMain
```

步骤 05 登录客户端。

zkCli.sh在zookeeper_home/bin目录下,可以配置ZooKeeper的环境变量。配置好环境变量之后,可以在任意的目录下执行zkCli.sh命令登录zkServer。

```
export ZOOKEEPER_HOME=/opt/zookeeper-3.6.2
export PATH=.:$PATH:$ZOOKEEPER_HOME/bin
```

登录ZooKeeper需要使用zkCli.sh命令。可以使用以下任意一种登录方式。

(1)登录本地ZooKeeper,直接输入zkCli.sh即可:

```
$zkCli.sh
[zk: localhost:2181(CONNECTED) 0]
```

(2)登录远程ZooKeeper服务器。

通过-server指定服务器的名称地址和端口:

```
$ zkCli.sh -server 192.168.56.101:2181
```

登录时,指定默认登录的目录:

```
$ zkCli.sh  -server  192.168.56.11:2181/zookeeper
```

5.3 基本客户端命令

1. 显示命令帮助

使用zkCli.sh登录ZooKeeper以后,可以直接使用help显示所有ZooKeeper的命令列表:

```
[zk: localhost:2181(CONNECTED) 0] help
ZooKeeper -server host:port -client-configuration properties-file cmd args
```

```
            addWatch [-m mode] path # optional mode is one of [PERSISTENT,
PERSISTENT_RECURSIVE] - default is PERSISTENT_RECURSIVE
            addauth scheme auth
            close
            config [-c] [-w] [-s]
            connect host:port
            create [-s] [-e] [-c] [-t ttl] path [data] [acl]
            delete [-v version] path
            deleteall path [-b batch size]
            delquota [-n|-b] path
            get [-s] [-w] path
            getAcl [-s] path
            getAllChildrenNumber path
            getEphemerals path
            history
            listquota path
            ls [-s] [-w] [-R] path
            printwatches on|off
            quit
            reconfig [-s] [-v version] [[-file path] | [-members
serverID=host:port1:port2;port3[,...]*]] | [-add
serverId=host:port1:port2;port3[,...]]* [-remove serverId[,...]*]
            redo cmdno
            removewatches path [-c|-d|-a] [-l]
            set [-s] [-v version] path data
            setAcl [-s] [-v version] [-R] path acl
            setquota -n|-b val path
            stat [-w] path
            sync path
            version
```

以下是一些具体的操作示例。

2. 查看znode节点

```
[zk: localhost:2181(CONNECTED) 1] ls /
[zookeeper]
```

或使用-R参数，显示一个节点树：

```
[zk: localhost:2181(CONNECTED) 2] ls / -R
/
/zookeeper
/zookeeper/config
/zookeeper/quota
```

使用ls -s显示某个节点的详细信息：

```
[zk: localhost:2181(CONNECTED) 3] ls -s /zookeeper
[config, quota]
cZxid = 0x0
ctime = Thu Jan 01 08:00:00 CST 1970
mZxid = 0x0
mtime = Thu Jan 01 08:00:00 CST 1970
pZxid = 0x0
cversion = -2
dataVersion = 0
```

```
aclVersion = 0
ephemeralOwner = 0x0
dataLength = 0
numChildren = 2
```

3. 创建一个znode节点

```
[zk: localhost:2181(CONNECTED) 4] create /test SomeValue
Created /test
```

创建多个子节点，如果父节点已经存在，则可以创建子节点，否则不能创建：

```
[zk: localhost:2181(CONNECTED) 5] create /test/subNode SomeValue
Created /test/subNode
```

使用-e参数添加一个临时节点，当客户端退出时，这种类型的节点将自动被删除。我们可以通过临时节点的这个特性来判断集群中的哪一个节点已经下线：

```
[zk: localhost:2181(CONNECTED) 6] create -e /test2 ThisIsTemporaryNode
Created /test2
```

4. 显示某个节点中的数据

```
[zk: localhost:2181(CONNECTED) 7] get /test
SomeValue
```

5. 删除某个节点

```
[zk: localhost:2181(CONNECTED) 9] delete /test/subNode
```

6. 删除节点及子节点

```
[zk: localhost:2181(CONNECTED) 9] deleteall /test
```

7. 设置节点的数据

```
[zk: localhost:2181(CONNECTED) 10] set /test newValue
[zk: localhost:2181(CONNECTED) 11] get /test
newValue
```

8. 创建一个观察者

首先在一个客户端A执行以下代码，注意后面的watch：

```
[zk: localhost:2181(CONNECTED) 12] stat /test watch
```

然后再打开一个客户端B，修改/test节点中的数据并执行：

```
[zk: localhost:2181(CONNECTED) 1] set /test NewData
```

此时，将会在客户端A看到以下代码输出：

```
[zk: localhost:2181(CONNECTED) 13]
WATCHER::
WatchedEvent state:SyncConnected type:NodeDataChanged path:/test
```

- state：状态，SyncConnected为连接状态。
- type：类型，NodeDataChanged为节点数据发生变化。

也可以使用ls -w观察一个目录的变化。例如：

```
ls -w /test
[zk: localhost:2181(CONNECTED) 0] ls -w /test
[]
```

如果向test下添加或是删除子节点，将会显示以下数据信息：

```
[zk: localhost:2181(CONNECTED) 1]
WATCHER::
WatchedEvent state:SyncConnected type:NodeChildrenChanged path:/test
```

5.4　Java 代码操作 ZooKeeper

使用Java代码操作ZooKeeper，首先需要添加ZooKeeper的依赖：

```xml
<dependency>
    <groupId>org.apache.zookeeper</groupId>
    <artifactId>zookeeper</artifactId>
    <version>3.6.2</version>
</dependency>
```

1. 读取节点中的数据

通过getData(..)函数，可以读取节点中的数据。示例代码如下：

代码 5.1　ReadData.java

```java
01 package org.hadoop.zk;
02 import org.apache.zookeeper.Watcher;
03 import org.apache.zookeeper.ZooKeeper;
04 import org.apache.zookeeper.data.Stat;
05 import java.util.concurrent.CountDownLatch;
06 public class ReadData {
07     public static void main(String[] args) throws Exception {
08         //定义一个线程计数器，用于线程通信
09         CountDownLatch countDownLatch = new CountDownLatch(1);
10         ZooKeeper zooKeeper = new ZooKeeper("server101:2181", 30000, (e) -> {
11             if (e.getState() == Watcher.Event.KeeperState.SyncConnected) {
12                 countDownLatch.countDown();
13             }
14         });
15         countDownLatch.await();//等待
16         byte[] bs = zooKeeper.getData("/test", null, new Stat());
17         String str = new String(bs);//将读取到的数据放到String对象中
18         System.out.println("读取到的数据是: " + str);
19         zooKeeper.close();
20     }
21 }
```

程序运行后，会在控制台上输出读取的数据。

2. 向已经存在的节点写入新的数据

通过setData(..)最多可以保存1MB的数据。因为Node并非文件系统中的目录,所以Node本身保存的数据是有限的。如代码5.2所示。

代码 5.2　WriteData.java

```java
01 package org.hadoop.zk;
02 import org.apache.zookeeper.Watcher;
03 import org.apache.zookeeper.ZooKeeper;
04 import org.apache.zookeeper.data.Stat;
05 import java.util.concurrent.CountDownLatch;
06 public class WriteData {
07     public static void main(String[] args) throws Exception {
08         CountDownLatch countDownLatch = new CountDownLatch(1);
09         ZooKeeper zooKeeper = new ZooKeeper("server101:2181", 30000, (e) -> {
10             if (e.getState() == Watcher.Event.KeeperState.SyncConnected) {
11                 countDownLatch.countDown();
12             }
13         });
14         countDownLatch.await();
15         Stat exists = zooKeeper.exists("/test", false);
16         if (exists != null) {
17             //最多可以保存1MB的数据,-1将会自动的增长
18             Stat stat = zooKeeper.setData("/test", "This Is New Data".getBytes(), -1);
19             System.out.println("数据版本号: " + stat.getVersion());
20         }
21         zooKeeper.close();
22     }
23 }
```

现在查询此节点中的数据,发现已经发生变化:

```
[zk: localhost:2181(CONNECTED) 1] get /test
This Is New Data
```

3. 创建新的节点

通过create方法可以创建一个新的节点,并可以指定节点的类型。如代码5.3所示。

代码 5.3　CreateNode.java

```java
01 package org.hadoop.zk;
02 import org.apache.zookeeper.CreateMode;
03 import org.apache.zookeeper.Watcher;
04 import org.apache.zookeeper.ZooDefs;
05 import org.apache.zookeeper.ZooKeeper;
06 import java.util.concurrent.CountDownLatch;
07 public class CreateNode {
08     public static void main(String[] args) throws Exception {
09         CountDownLatch countDownLatch = new CountDownLatch(1);
10         ZooKeeper zooKeeper = new ZooKeeper("server101:2181",
11                 30000, (e) -> {
12             if (e.getState() == Watcher.Event.KeeperState.SyncConnected) {
```

```
13              countDownLatch.countDown();
14          }
15      });
16      countDownLatch.await();
17      //OPEN_ACL_UNSAFE开放安全,PERSISTENT持久化节点类型
18      String path = zooKeeper.create("/test2", "New ZNode".getBytes(),
19              ZooDefs.Ids.OPEN_ACL_UNSAFE,
20              CreateMode.PERSISTENT);
21      System.out.println("创建成功: " + path);
22      zooKeeper.close();
23  }
24 }
```

4. 删除节点

通过delete函数,可以删除一个节点。如代码5.4所示。

代码 5.4　DeleteNode.java

```
01 package org.hadoop.zk;
02 import org.apache.zookeeper.Watcher;
03 import org.apache.zookeeper.ZooKeeper;
04 import java.util.concurrent.CountDownLatch;
05 public class DeleteNode {
06     public static void main(String[] args) throws Exception {
07         CountDownLatch countDownLatch = new CountDownLatch(1);
08         ZooKeeper zooKeeper = new ZooKeeper("taoqiu:2181",
09                 30000, (e) -> {
10             if (e.getState() == Watcher.Event.KeeperState.SyncConnected) {
11                 countDownLatch.countDown();
12             }
13         });
14         countDownLatch.await();
15         //-1为删除所有的版本
16         zooKeeper.delete("/test2", -1);
17         zooKeeper.close();
18     }
19 }
```

5. 创建观察者

通过开发观察者,可以监控到节点中数据的变化。如代码5.5所示。

代码 5.5　NodeWatch.java

```
01 package org.hadoop.zk;
02 import org.apache.zookeeper.WatchedEvent;
03 import org.apache.zookeeper.Watcher;
04 import org.apache.zookeeper.ZooKeeper;
05 import org.apache.zookeeper.data.Stat;
06 import java.util.concurrent.CountDownLatch;
07 public class NodeWatch {
08     public static void main(String[] args) throws Exception {
09         CountDownLatch countDownLatch = new CountDownLatch(1);
```

```
10      ZooKeeper zooKeeper = new ZooKeeper("server101:2181", 30000, (e)->{
11          if(e.getState()== Watcher.Event.KeeperState.SyncConnected){
12              countDownLatch.countDown();
13          }
14      });
15      countDownLatch.await();
16      //添加观察者，只观察一次，如果希望循环观察，可以在接收到观察的信息以后再次添加观察者
17      //以下将观察/test节点的数据变化
18      Watcher watcher = new Watcher() {
19          @Override
20          public void process(WatchedEvent event) {
21              System.out.println(event.getType()+","+event.getState());
22          }
23      };
24      zooKeeper.getData("/test", watcher,new Stat());
25      Thread.sleep(Integer.MAX_VALUE);
26      zooKeeper.close();
27  }
28 }
```

现在启动上面的代码，并使用命令行修改/test节点中的数据：

```
set /test NewData
```

Java代码编译器控制台将会显示以下信息，即数据发生变化：

```
NodeDataChanged,SyncConnected
```

6. 将整个文本数据保存到节点中

一个节点最多可以保存1MB的数据。ZooKeeper保存数据只能通过节点的datas来保存，因此不能保存太大的数据，对于小文本数据，如solr的配置文件数据，都是通过节点的datas来保存的。代码5.6中将Java代码保存到同名的Java节点中。

代码 5.6　SaveDatas.java

```
01 package org.hadoop.zk;
02 import org.apache.zookeeper.*;
03 import java.io.File;
04 import java.nio.file.Files;
05 import java.util.concurrent.CountDownLatch;
06 public class SaveDatas implements Watcher {
07     public static void main(String[] args) throws Exception {
08         new SaveDatas();
09     }
10     CountDownLatch countDownLatch = new CountDownLatch(1);
11     public SaveDatas() throws Exception {
12         ZooKeeper zk = new ZooKeeper("server101:2181", 2000, this);
13         countDownLatch.await();
14         File file = new
   File("./chapter07/src/test/java/org/hadoop/zk/SaveDatas.java");
15         //获取这个文件的所有数据
16         byte[] bs = Files.readAllBytes(file.toPath());
17         //在创建目录的同时，将数据保存到这个节点中
```

```
18            zk.create("/test/" + file.getName(), bs, ZooDefs.Ids.OPEN_ACL_UNSAFE,
   CreateMode.PERSISTENT);
19            zk.close();
20        }
21        @Override
22        public void process(WatchedEvent event) {
23            if (event.getState() == Event.KeeperState.SyncConnected && event.getType() ==
   Event.EventType.None) {
24                countDownLatch.countDown();
25            }
26        }
27 }
```

查看执行结果:

```
[zk: localhost:2181(CONNECTED) 5] ls /test
[SaveDatas.java, a, subDir]
```

也可以通过SaveDatas.java文件查看里面的数据,即可看到此文件的代码已经上传到zk。

7. 从指定的节点中获取数据

可以将指定的ZooKeeper节点的数据保存下载到本地。代码5.7所示是一个可以运行的Java代码。

代码 5.7　ZkDownload.java

```
01 package org.hadoop.zk;
02 import org.apache.zookeeper.WatchedEvent;
03 import org.apache.zookeeper.Watcher;
04 import org.apache.zookeeper.ZooKeeper;
05 import org.apache.zookeeper.data.Stat;
06 import java.io.FileOutputStream;
07 import java.io.OutputStream;
08 import java.util.concurrent.CountDownLatch;
09 public class ZkDownload {
10     public static void main(String[] args) throws Exception {
11         if (args.length < 2) {
12             System.out.println("用法: java cn.zk.ZkDownload <zk集群地址> <zk节点路径> [保
   存文件名]");
13             return;
14         }
15         String zkServer = args[0];
16         String path = args[1];
17         String dest = null;
18         if (args.length >= 3) {//如果传递了第三个参数,即保存的文件
19             dest = args[2];
20         }
21         CountDownLatch countDownLatch = new CountDownLatch(1);
22         ZooKeeper zooKeeper = new ZooKeeper(zkServer, 3000, new Watcher() {
23             @Override
24             public void process(WatchedEvent event) {
25                 countDownLatch.countDown();
26             }
27         });
28         countDownLatch.await();
```

```
29          Stat stat = zooKeeper.exists(path, null);
30          if (stat == null) {
31              throw new RuntimeException("节点:" + path + "不存在!");
32          }
33          byte[] bs = zooKeeper.getData(path, false, stat);
34          if (dest != null) {
35              path = dest;
36          } else {
37              path = path.substring(path.lastIndexOf("/") + 1);//只取出最后一个节点的名称
38          }
39          OutputStream out = new FileOutputStream(path);
40          out.write(bs);
41          out.close();
42          System.out.println("下载完成");
43      }
44  }
```

现在我们将它上传到Linux服务器上,并编译执行(注意classpath的值里面有一个点):

```
$ javac -classpath .:/app/zookeeper/lib/*  \
-d . ZkDownload.java
```

下载指定的节点的数据:

```
$ java -classpath .:/app/zookeeper/lib/* org.hadoop.zk.ZkDownload localhost:2181 /test a.txt
```

执行完成以后,查看a.txt文件中的内容,就是/test节点中的数据。

5.5　ZooKeeper 集群安装

想要安装ZooKeeper集群,首先需要在参与集群的多台机器上都安装ZooKeeper,为了便于记忆,可以将ZooKeeper安装到相同的目录下,如/app目录下。

在每一个ZooKeeper的dataDir目录下,创建一个myid文件,里面保存的是当前ZooKeeper节点的id。id不一定要从1开始。本示例中的id根据本机IP地址最后一段作为id,以便于记忆。

步骤01 配置zoo.cfg。

准备三台CentOS7主机,且关闭防火墙。然后修改$ZOOKEEPER_HOME/conf/zoo.cfg,在文件最后追加以下配置:

```
#配置ZooKeeper数据保存目录
dataDir=/app/datas/zk
```

配置ZooKeeper集群:

```
server.101=192.168.56.11:2888:3888
server.102=192.168.56.12:2888:3888
server.103=192.168.56.13:2888:3888
```

步骤02 使用scp将文件发送到其他两台计算机上。

```
$ scp -r zookeeper-3.6.2  server102:/app/
```

```
$ scp -r zookeeper-3.6.2 server103:/app/
```

然后修改每一个dataDir目录下的 myid文件。在server101主机上的myid中添加101（这里是为了方便记忆才取此id的，此id等于IP地址，可以从1开始），即：

```
echo 101 > /app/datas/zk/myid
```

以此类推。

步骤 03 分别启动三台主机。

启动zkServer：

```
$ ./zkServer.sh start
```

查看状态，使用status参数检查服务器状态：

```
[hadoop@server101 bin]$ ./zkServer.sh status
Mode: leader    #这是Leader，其他两台为Follower
```

现在就可以进行同步测试了。在一台机器上进行操作，并查看其他两台的同步情况。

步骤 04 测试操作。

登录客户端：

```
$ zkCli.sh
```

显示当前所有目录：

```
[hadoop: localhost 2181] ls /
[zookeeper]
```

创建一个新的目录，且写入数据：

```
[hadoop: localhost : 2181 ] create /test TestData
```

再次显示当前根目录下的所有数据，也可以登录其他主机，如果查看到相同的结果，即表示已经同步成功。

```
[hadoop: localhost : 2181 ] ls /
[test, zookeeper]
```

ZooKeeper的命令很多，可以使用help查看所有可使用的命令：

```
[zhadoopk: localhost : 2181] help
```

通过上面的配置可以看出，ZooKeeper的集群配置相对比较简单。只要配置zoo.cfg并指定所有服务节点，然后在每一个节点的data目录下，将当前id写入myid文件中即可。

5.6 znode 节点类型

znode的节点类型分为持久节点、顺序节点和临时节点。默认创建的节点都是持久节点，使用-s参数可以创建一个顺序节点，使用-e参数可以创建临时节点。

创建顺序节点：

```
[hadoop: localhost:2181(CONNECTED) 13] create -s /t ""
Created /t0000000004
```

创建临时节点，客户端退出时会自动被删除：

```
[hadoop: localhost:2181(CONNECTED) 2] create -e /tt ""
Created /tt
```

5.7 观察节点

观察者Oberserver可以观察到节点的变化。

```
stat path [watch]
ls path [watch]
ls2 path [watch]
get path [watch]
```

上面的命令后面都有一个watch参数，即表示可以在命令行观察节点的变化。

首先通过某个设置在后面直接添加watch：

```
[hadoop: localhost:2181(CONNECTED) 23] ls /one watch
[]
```

然后在另一个客户端添加一个子节点：

```
[hadoop: localhost:2181(CONNECTED) 4] create /one/1 Some
Created /one/1
```

现在可以观察到节点显示的数据：

```
[hadoop: localhost:2181(CONNECTED) 24]
WATCHER::
WatchedEvent state:SyncConnected type:NodeChildrenChanged path:/one
```

5.8 配置 Hadoop 高可用集群

Hadoop 2.x已经发布了稳定版本，增加了很多特性，比如HDFS HA、YARN都可以配置HA（高可用）。Hadoop 3.x相对于2.x则具有一些更高级的特性，兼容性也更好。对于Hadoop 2.x版本，一个namenode service最多只能有两个NameNode，而到了3.0版本，一个namenodeservice可以有三个NameNode或更多。以下是官方说明：

HA的NameNode最小数量是两个，但可以配置更多。由于通信开销，建议不要超过5个，一般推荐3个NameNode。

配置Hadoop高可用将会启动一些新的进程，它们是：

- ZKFC（DFSZKFailoverController）：Hadoop进程，ZooKeeper通过与ZKFC通信获取NameNode的活动状态。如果ZKFC认为NameNode已经宕机，将会通过ZooKeeper开启新的选举程序，选择出新的主NameNode。
- QJM（Quorum Journal Manager）：用于同步主NameNode的日志数据，将其保存到备份NameNode中。在高可用情况下，将不再使用SecondaryNameNode程序，多个NameNode之间互相备份。QJM负责它们的通信数据。

由于配置Hadoop高可用集群比较复杂，因此我们需要做好配置前的规划。
以下是具体的配置规划：

- 配置从一台主机到其他主机的SSH免密码登录，主要是执行start-dfs.sh和start-yarn.sh的主机到其他主机的SSH免密码登录。
- 关闭所有主机的防火墙，仅非生产环境。
- 配置所有主机的静态地址和hosts文件。
- 所有主机上安装JDK 1.8+，并配置JAVA环境变量。
- 至少在三台主机上安装ZooKeeper，并启动ZooKeeper集群。
- 在一台主机上配好Hadoop的所有配置文件，并通过scp分发到所有主机的相同目录下。
- 所有主机配置Hadoop环境变量。
- 启动JournalNode。
- 在某台配置了NameNode的主机上格式化NameNode。
- 将格式化后的目录复制到其他的主机（注意只是指配置了NameNode的主机）。
- 格式化ZKFC。
- 启动HDFS，启动YARN。

步骤01 配置计划表（见表5-4）。

表 5-4 Hadoop 高可用集群配置计划表

IP/主机名	软件	进程
192.168.56.101 server101	JDK 1.8+ ZooKeeper 3.6.2 Hadoop 3.2.2	QuorumPeerMain NameNode ZKFC QJM ResourceManager NodeManager DataNode
192.168.56.102 server102	JDK 1.8+ ZooKeeper 3.6.2 Hadoop 3.2.2	QuorumPeerMain NameNode ZKFC QJM ResourceManager NodeManager DataNode

(续表)

IP/主机名	软件	进程
192.168.56.103 server103	JDK 1.8+ ZooKeeper 3.6.2 Hadoop 3.2.2	QuourmPeerMain NameNode ZKFC QJM ResourceManager NodeManager DataNode

步骤02 前期准备。

所有主机关闭防火墙：

```
$ sudo sytemctl stop firewalld
$ sudo systemctl disable firewalld
```

所有主机安装JDK 1.8+，并配置环境变量：

```
$ sudo tar -zxvf ~/jdk1.8-281.tag.gz -C /usr/java/
$ sudo vim /etc/provfile
export JAVA_HOHE=/usr/java/jdk1.8-281
exoort PATH=$PATH:$JAVA_HOME/bin
```

所有主机设置静态IP地址，修改主机名称。需要分别修改三台主机，以下以一台主机为例：

```
$ sudo vim /etc/sysconfig/network-scripts/ifcfg-enp0s8
IPADDR=192.168.56.101
$ sudo systemctl hostnamectl set-hostname server101
$ sudo vim /etc/hosts
192.168.56.101 server101
192.168.56.102 server102
192.168.56.103 server103
```

设置所有主机selinux=disabled：

```
$ sudo vim /etc/selinux/config
selinux=disabled
```

安装好ZooKeeper集群，并启动。请参考5.3节。

步骤03 配置hadoop-env.sh文件。

在hadoop-env.sh文件中，添加JAVA_HOME环境变量：

```
export JAVA_HOME=/usr/local/java/jdk1.8.0_281
```

步骤04 配置core-site.xml文件。

与之前的配置一样，配置Hadoop的core-site.xml文件，但请注意，之前配置fs.defaultFS的值为hdfs://server101:8020，在集群模式下，其值需要配置为hdfs://cluster。其中cluster为任意设置的名称，请牢记此名称，后台将会使用此名称来配置NameService。

```
01 <configuration>
02     <property>
03         <name>fs.defaultFS</name>
```

```
04        <value>hdfs://cluster</value>
05    </property>
06    <property>
07        <name>hadoop.tmp.dir</name>
08        <value>/app/datas/hadoop</value>
09    </property>
10    <property>
11        <name>ha.zookeeper.quorum</name>
12        <value>server101:2181,server102:2181,server103:2181</value>
13    </property>
14 </configuration>
```

配置说明：

第04行：指定集群名称，其中cluster可以是任意的名称，但请牢记此名称，后面将会用到。

第08行：与之前一样，配置NameNode保存数据的本地目录，此目录使用hdfs namenode -format格式化后将会有数据。将格式化完成的目录通过scp复制到其他NameNode节点的相同目录下，这样做的目的是为了保证Hadoop集群ID的唯一。

第12行：指定ZooKeeper的集群地址。由于我们主机不是很多，所以这里ZooKeeper的地址很多与Hadoop存储计算结果相同，在正式的生产环境下，此ZooKeeper节点应该有独立的主机。

步骤 05 配置hdfs-site.xml文件。

这里的配置信息比较多，请注意观察。hdfs-site.xml文件配置与之前一样，不过在集群环境下，还需要配置集群所对应的NameService。NameService是虚拟集群服务的名称。由于配置比较多，所以我们将每一个配置的具体含义都直接添加到了配置项上进行说明。

```
<configuration>
    <!--指定HDFS的NameService为cluster，需要和core-site.xml中的保持一致 -->
    <property>
        <name>dfs.nameservices</name>
        <value>cluster</value>
    </property>
    <!-- cluster下面有多个NameNode,分别取名为nn1、nn2、nn3,也可以取其他名称。注意,NameNode后缀为cluster,即之前配置的名称-->
    <property>
        <name>dfs.ha.namenodes.cluster</name>
        <value>nn1,nn2,nn3</value>
    </property>
    <!--配置每一个NameNode的rpc通信地址-->
    <property>
        <name>dfs.namenode.rpc-address.cluster.nn1</name>
        <value>server101:8020</value>
    </property>
    <property>
        <name>dfs.namenode.rpc-address.cluster.nn2</name>
        <value>server102:8020</value>
    </property>
    <property>
        <name>dfs.namenode.rpc-address.cluster.nn3</name>
        <value>server103:8020</value>
    </property>
    <!--配置每一个NameNode的Web http地址-->
```

```xml
<property>
    <name>dfs.namenode.http-address.cluster.nn1</name>
    <value>server101:9870</value>
</property>
<property>
    <name>dfs.namenode.http-address.cluster.nn2</name>
    <value>server102:9870</value>
</property>
<property>
    <name>dfs.namenode.http-address.cluster.nn3</name>
    <value>server103:9870</value>
</property>
<!--配置QJM的地址-->
<property>
    <name>dfs.namenode.shared.edits.dir</name>
    <value>
    qjournal://server101:8485;server102:8485;server103:8485/cluster
    </value>
</property>
<!--配置QJM日志的目录-->
<property>
    <name>dfs.journalnode.edits.dir</name>
    <value>/app/datas/hadoop/qjm</value>
</property>
<!--配置为自动切换功能打开,需要在core-site.xml文件中配置ZK地址-->
<property>
    <name>dfs.ha.automatic-failover.enabled</name>
    <value>true</value>
</property>
<property>
    <name>dfs.client.failover.proxy.provider.cluster</name>
    <value>
org.apache.hadoop.hdfs.server.namenode.ha.ConfiguredFailoverProxyProvider
    </value>
</property>
<!--配置自动切换的方式-->
<property>
    <name>dfs.ha.fencing.methods</name>
    <value>
        sshfence
        shell(/bin/true)
    </value>
</property>
<!--配置SSH key,注意根据不同的用户名修改目录-->
<property>
    <name>dfs.ha.fencing.ssh.private-key-files</name>
    <value>/home/hadoop/.ssh/id_rsa</value>
</property>
<property>
    <name>dfs.permissions.enabled</name>
    <value>false</value>
</property>
<!-- 配置sshfence隔离机制超时时间 -->
<property>
    <name>dfs.ha.fencing.ssh.connect-timeout</name>
    <value>30000</value>
```

```
        </property>
</configuration>
```

步骤 06 配置mapred-site.xml。

配置MapReduce与之前的配置完全一样，配置调试模式为YARN即可：

```xml
<configuration>
    <!-- 指定mr框架为yarn方式 -->
    <property>
        <name>mapreduce.framework.name</name>
        <value>yarn</value>
    </property>
</configuration>
```

步骤 07 配置yarn-site.xml。

配置yarn-site.xml文件与配置hdfs-site.xml一样，需要指定集群的配置环境。为了方便阅读，我们直接将配置说明添加到了配置功能之上进行说明。

```xml
<configuration>
    <!--配置RM高可靠启用-->
    <property>
        <name>yarn.resourcemanager.ha.enabled</name>
        <value>true</value>
    </property>
    <!-- 给resource manager取一个任意的名称-->
    <property>
        <name>yarn.resourcemanager.cluster-id</name>
        <value>cluster1</value>
    </property>
    <!--配置Resourcemanager个数，Hadoop 3以后可以为3个以上，值部分为任意名称-->
    <property>
        <name>yarn.resourcemanager.ha.rm-ids</name>
        <value>rm1,rm2,rm3</value>
    </property>
    <!--以下配置每一个RM的地址-->
    <property>
        <name>yarn.resourcemanager.hostname.rm1</name>
        <value>server101</value>
    </property>
    <property>
        <name>yarn.resourcemanager.hostname.rm2</name>
        <value>server102</value>
    </property>
    <property>
        <name>yarn.resourcemanager.hostname.rm3</name>
        <value>server103</value>
    </property>
    <!--配置每一个RM的http地址-->
    <property>
        <name>yarn.resourcemanager.webapp.address.rm1</name>
        <value>server101:8088</value>
    </property>
    <property>
        <name>yarn.resourcemanager.webapp.address.rm2</name>
        <value>server102:8088</value>
```

```xml
        </property>
        <property>
            <name>yarn.resourcemanager.webapp.address.rm3</name>
            <value>server103:8088</value>
        </property>
        <!--配置ZooKeeper地址-->
        <property>
            <name>yarn.resourcemanager.zk-address</name>
            <value>server101:2181,server102:2181,server103:2181</value>
        </property>
        <property>
            <name>yarn.nodemanager.aux-services</name>
            <value>mapreduce_shuffle</value>
        </property>
        <!--Hadoop 3里面必须添加的classpath-->
        <property>
            <name>yarn.application.classpath</name>
            <value>
                请自行将Hadoop classpath执行结果添加到这儿
            </value>
        </property>
</configuration>
```

步骤08 配置workers文件。

workers是指定DataNode节点的位置。在里面添加主机的名称或是IP地址均可，一行一个节点：

```
server101
server102
server103
```

步骤09 配置Hadoop的环境变量。

在server101主机上配置Hadoop环境变量，并通过scp将profile文件分发到其他主机的相同目录下：

```
$sudo vim /etc/profile
export HADOOP_HOME=/app/hadoop-3.2.2
export PATH=$PATH:$HADOOP_HOME/bin
$ sudo scp /etc/profile root@server102:/etc/
$ sudo scp /etc/profile root@server103:/etc/
```

让环境变量生效，执行以下命令：

```
$source /etc/profile
```

步骤10 复制文件到其他主机。

将配置好的Hadoop目录和Hadoop配置文件，复制到其他主机相同的目录下（使用scp命令）。由于share目录下的doc里面都是文档，可以删除这个目录，以加快复制速度。

删除doc目录：

```
$ rm -rf /app/hadoop-3.2.2/share/doc
```

复制文件：

```
$ scp -r /app/hadoop-3.2.2  server102:/app/
$ scp -r /app/hadoop-3.2.2  server103:/app/
```

步骤 11 启动journalnode。

分别在在server101、server102、server103主机上执行启动JournalNode的工作:

```
$ /app/hadoop-3.2.2/bin/hadoop-daemon.sh start journalnode
```

步骤 12 格式化HDFS。

在server101主机上执行格式化NameNode的命令:

```
$ hdfs namenode -format
```

格式化后，HDFS会根据core-site.xml中的hadoop.tmp.dir配置生成个文件，然后将这个文件使用scp复制到server102、server103的相同目录下。因为都是NameNode节点，必须拥有相同的数据文件。格式化成功的标志是在输出的日志中查看是否存在以下语句:

```
Storage directory /opt/hadoop_tmp_dir/dfs/name has been successfully formatted
```

现在将格式化后的HDFS目录，复制到weric12主机上的相同目录下:

```
$ scp -r /app/datas/hadoop  server102:/app/datas/
$ scp -r /app/datas/hadoop  server103:/app/datas/
```

步骤 13 格式化ZKFC。

在server101上执行格式化ZKFC的代码如下:

```
$ hdfs zkfc -formatZK
Successfully created /hadoop-ha/cluster in ZK.
```

在格式化完成以后，通过zkCli.sh登录ZooKeeper并查看目录列表，结果将显示一个hadoop-ha的目录，表示初始化成功:

```
[zk: localhost:2181(CONNECTED) 0] ls /
[zookeeper, hadoop-ha]
```

步骤 14 启动HDFS。

在server101上启动HDFS即NameNode，此命令会根据配置文件，同时启动server102、server103上的NameNode:

```
$ /app/hadoop-3.2.2/sbin/start-dfs.sh
```

现在就可以通过haadmin命令查看整个集群的情况，此命令会输出集群中NameNode节点的活动状态，其中active为活动节点，standby为备份节点:

```
$ hdfs haadmin -getAllServiceState
server101:8020      active
server102:8020      standby
server103:8020      standby
```

步骤 15 启动YARN。

在server101上执行启动YARN的命令，其会根据配置信息，同时启动server102和server103的ResouceManager进程:

```
$ /app/hadoop-3.2.2/sbin/start-yarn.sh
```

第 5 章 ZooKeeper与高可用集群实战 | 101

在启动完成以后，根据之前的配置列表，分别检查每一个主机上的服务是否都已经启动。如果没有，请查看日志错误排查问题。

启动完成以后，查看ResourceManager的集群状态，可以使用rmadmin命令。与NameNode一样，active表示活动节点，standby为准备节点：

```
$ yarn rmadmin -getAllServiceState
server101:8033      active
server102:8033      standby
server103:8033      standby
```

步骤16 验证高可用。

通过浏览器访问一下地址，可以查看每一台HDFS的信息。server102、server103由于是准备节点，所以显示的状态为standby。访问http://server102:9870，如图5-3所示。

图5-3　访问http://server102:9870

访问http://server103:9870，如图5-4所示。

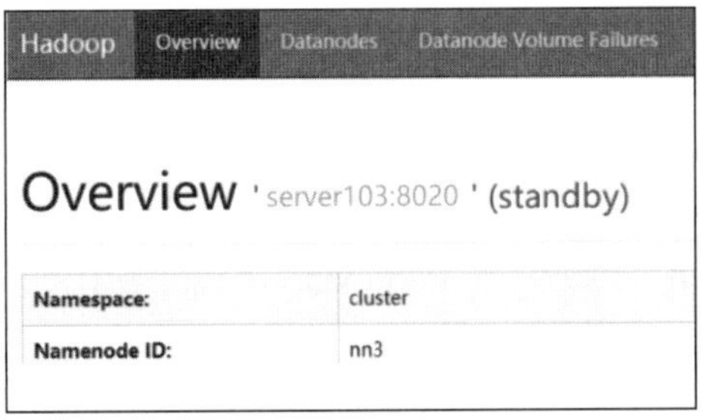

图5-4　访问http://server103:9870

从图5-4可以看出，当前server101的NameNode为active。

访问http://server101:9870，如图5-5所示。

图5-5　访问http://server101:9870

也可以通过以下命令，检查NameNode和ResourceManager的状态：

```
$ hdfs haadmin -getServiceState nn1
active
$ hdfs haadmin -getServiceState nn2
standby
$ yarn rmadmin -getServiceState rm1
active
$ yarn rmadmin -getServiceState rm2
standby
```

现在让我们挂掉active的NameNode，即挂掉nn1：

```
$ kill -9 <pid of NN>
```

然后再检查状态，这时server102上的NameNode变成了active：

```
$ hdfs haadmin -getServiceState nn2
Active
```

也可以通过haadmin查看节点状态：

```
$ hdfs haadmin -getAllServiceState
server101    failed to connected
server102    active
server103    standby
```

手动启动那个挂掉的NameNode，即nn1，然后再检查状态，它已经成为standby了。

```
$ ./hadoop-daemon.sh start namenode
$ hdfs haadmin -getServiceState nn1
standby
```

使用同样的方式，可以验证ResourceManager是否可以自动实现容灾切换。

注意：

（1）在集群完成以后，建议执行一个MapReduce测试，如本书开始的WordCount示例。

（2）Hadoop高可用集群每一次启动相对比较麻烦。但配置成功以后，下次启动就比较简单了。对于上面的示例而言，再次启动只要在server101主机上执行./start-dfs.sh和./start-yarn.sh即可。

5.9 用 Java 代码操作集群

用Java客户端操作集群开发HDFS，必须指定NameService的配置信息。代码5.8用来显示HDFS上的文件和目录。

代码 5.8 HdfsAccess.java

```
01 package org.hadoop.ha;
02 import org.apache.hadoop.conf.Configuration;
03 import org.apache.hadoop.conf.Configured;
```

```java
04  import org.apache.hadoop.fs.FileStatus;
05  import org.apache.hadoop.fs.FileSystem;
06  import org.apache.hadoop.fs.Path;
07  import org.apache.hadoop.util.Tool;
08  import org.apache.hadoop.util.ToolRunner;
09  public class HdfsAccess extends Configured implements Tool {
10      @Override
11      public int run(String[] args) throws Exception {
12          System.setProperty("HADOOP_USER_NAME", "hadoop");
13          Configuration conf = getConf();
14          conf.set("fs.defaultFS", "hdfs://cluster");
15          conf.set("dfs.nameservices", "cluster");
16          conf.set("dfs.ha.namenodes.cluster", "nn1,nn2,nn3");
17          conf.set("dfs.namenode.rpc-address.cluster.nn1", "server101:8020");
18          conf.set("dfs.namenode.rpc-address.cluster.nn2", "server102:8020");
19          conf.set("dfs.namenode.rpc-address.cluster.nn3", "server103:8020");
20          conf.set("dfs.client.failover.proxy.provider.cluster",
21  "org.apache.hadoop.hdfs.server.namenode.ha.ConfiguredFailoverProxyProvider");
22          FileSystem fs = FileSystem.get(conf);
23          FileStatus[] statuses = fs.listStatus(new Path("/"));
24          for (FileStatus f : statuses) {
25              System.err.println(f.getPath().toString());
26          }
27          fs.close();
28          return 0;
29      }
30      public static void main(String[] args) throws Exception {
31          int code = ToolRunner.run(new HdfsAccess(), args);
32          System.exit(code);
33      }
34  }
```

代码5.9保存一个文件到高可用集群。

代码 5.9 HASaveHdfs.java

```java
01  package org.hadoop.ha;
02  import org.apache.hadoop.conf.Configuration;
03  import org.apache.hadoop.conf.Configured;
04  import org.apache.hadoop.fs.FileSystem;
05  import org.apache.hadoop.fs.Path;
06  import org.apache.hadoop.util.Tool;
07  import org.apache.hadoop.util.ToolRunner;
08  import java.io.OutputStream;
09  public class HASaveHdfs extends Configured implements Tool {
10      @Override
11      public int run(String[] args) throws Exception {
12          System.setProperty("HADOOP_USER_NAME", "hadoop");
13          Configuration conf = getConf();
14          conf.set("fs.defaultFS", "hdfs://cluster");
15          conf.set("dfs.nameservices", "cluster");
16          conf.set("dfs.ha.namenodes.cluster", "nn1,nn2,nn3");
17          conf.set("dfs.namenode.rpc-address.cluster.nn1", "server101:8020");
18          conf.set("dfs.namenode.rpc-address.cluster.nn2", "server102:8020");
19          conf.set("dfs.namenode.rpc-address.cluster.nn3", "server103:8020");
```

```
20          conf.set("dfs.client.failover.proxy.provider.cluster",
21  "org.apache.hadoop.hdfs.server.namenode.ha. ConfiguredFailoverProxyProvider");
22          FileSystem fs = FileSystem.get(conf);
23          OutputStream out = fs.create(new Path("/test/c.txt"));
24          out.write("Hello中文".getBytes());
25          out.close();
26          fs.close();
27          return 0;
28      }
29      public static void main(String[] args) throws Exception {
30          int code = ToolRunner.run(new HASaveHdfs(), args);
31          System.exit(code);
32      }
33  }
```

至此，你已经可以部署Hadoop高可用集群了，同时也学会了如何通过Java代码访问高可用集群。

5.10 小　　结

- ZooKeeper的配置。
- ZooKeeper Java编程接口。
- ZooKeeper集群配置。
- Hadoop借助ZooKeeper实现HA高可用。
- ZKFC用于时时向ZooKeeper汇报NameNode的状态，一旦NameNode不可用，就会自动进行切换。
- JournalNode用于时时同步两个NameNode之间的日志数据。
- 可以使用hdfs haadmin命令查看NameNode的集群状态。可以使用yarn rmadmin命令查看ResourceManager的集群状态。

第 6 章

Hive数据仓库实战

主要内容：

- Hive的体系结构和特点。
- Hive的命令。
- Hive视图。
- Hive表分区。
- Hive UDF编程。
- Hive的metastore server。

Hive是基于Hadoop的一个数据仓库，可以将结构化的数据文件映射为一张数据库表，并提供SQL查询功能。Hive在执行SQL时会将SQL语句转换为MapReduce任务运行。其优点是学习成本低，可以通过类似于SQL的语句快速实现MapReduce的开发，不必开发专门的MapReduce应用程序，十分适合离线数据的统计分析。

Hive是建立在Hadoop上的数据仓库基础构架。它提供了一系列的工具，可以用来进行数据提取－转化－加载（ETL），这是一种可以存储、查询和分析存储在Hadoop中的大规模数据的机制。Hive定义的类似于SQL的查询语言称为HQL，它允许熟悉SQL的用户查询数据。同时，这个语言也允许熟悉MapReduce开发者开发自定义的Mapper和Reducer，以用来处理内建的Mapper和Reducer无法完成的复杂的分析工作。

Hive3运行在Hadoop 3之上。以下是官方的发布说明，其中约定了Hive的什么版本可以运行在Hadoop的什么版本之上。

```
17 January 2021: release 2.3.8 available
This release works with Hadoop 2.x.y You can look at the complete JIRA change log
for this release.
26 August 2019: release 3.1.2 available
This release works with Hadoop 3.x.y. You can look at the complete JIRA change log
for this release.
```

Hive的特点如下：

- 对仓库中的数据进行分析和计算。
- 建立在Hadoop之上。
- 一次写入，多次读取。
- Hive是SQL语句分析引擎，将SQL语句转换成MapReduce并在Hadoop上执行。
- Hive表对应HDFS的文件夹。
- Hive的数据对应的是HDFS的文件。

Hive体系结构如图6-1所示。

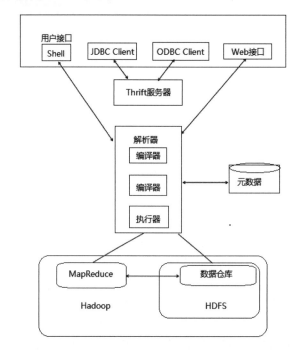

图6-1　Hive的体系结构

Hive的主要组成部分，分为以下几个部分：

1. 用户接口

用户接口主要有三个：CLI、Client和WUI。其中最常用的是CLI，CLI启动的时候，会同时启动一个Hive副本；Client是Hive的客户端，用户连接至Hive Server，在启动Client模式的时候，需要指出Hive Server所在节点，并且在该节点启动 Hive Server；WUI是通过浏览器访问Hive。

2. 元数据存储

Hive将元数据存储在数据库中，如MySQL、Derby。Hive中的元数据包括表的名字、表的列和分区及其属性、表的属性（是否为外部表等）、表的数据所在目录等。

目前已经支持的数据库如表6-1所示。

表 6-1　目前已经支持的数据库

支持的数据库	最小支持版本
MySQL	5.6.17
Postgres	9.1.13
Oracle	11g
MS SQL Server	2008 R2

3. 解释器、编译器、优化器、执行器

解释器、编译器、优化器和执行器的主要工作是对HQL查询语句进行词法分析、语法分析、编译、优化以及查询计划的生成。生成的查询计划存储在HDFS中，并随后由MapReduce调用执行。

Hive的数据存储在HDFS中，执行Hive的大部分SQL将会引发MapReduce的执行。

6.1　Hive3 的安装配置

到本书完成时，Hive已经更新到3.1.2版本。使用Hive的前置条件是启动HDFS，启动YARN。如果仅有一台服务器，并且已经安装了Hadoop伪分布式及ZooKeeper即可以开始Hive的学习。

步骤01 下载Hive3。

下载Hive地址为：https://mirrors.bfsu.edu.cn/apache/hive/hive-3.1.2/apache-hive-3.1.2-bin.tar.gz。

步骤02 解压。

可以将Hive解压到任意目录下，此处为了方便操作，将Hive解压到安装集群的Hadoop相同目录下：

```
$ tar -zxvf ~/apache-hive-3.1.2-bin.tar.gz -C /app/
```

步骤03 初始化数据库。

Hive的运行同样需要元数据的支持，Hive元数据可以保存到关系数据库中，默认的情况下保存到Derby数据库中。如果需要快速执行，可以直接选择使用Derby数据库。使用Derby数据库的缺点是，不能同时使用两个Hive CLI共同操作Hive数据仓库，因为Derby这种类型的数据库，不支持多用户同时登录。所以，我们需要将元数据保存到类似于MySQL的关系数据库中，这将在后面的章节详细讲解。

（1）使用Derby数据库保存元数据

默认情况下，Hive使用Derby数据库作为元数据管理数据库。使用Derby数据库时，会在当前目录下创建metastore_db目录，即为Derby数据库的目录。与Hive2和Hive1不同，Hive3在使用Derby数据库之前，必须先执行初始化命令，一个Derby数据库只允许一个Hive客户端登录。

以下初始化Derby数据库（请确定已经启动HDFS）：

```
$ /app/hive-3.1.2/bin/schematool -dbType derby -initSchema
```

成功后显示：

```
Initialization script completed
schemaTool completed
```

初始化成功以后,就可以使用hive命令登录Hive客户端:

```
$/app/hive-3.1.2/bin/hive
Hive>
```

如果显示了Hive>则表示登录Hive的命令行成功,此时就可以使用Hive SQL操作数据了。Hive SQL使用了类似于MySQL的SQL语法,因此有SQL编程经验的读者,可以快速上手。

显示当前有多少数据库:

```
Hive>show databases;
Default
```

创建一个数据库,会在HDFS的user/hive/warehouse/下创建一个xx.db的数据库目录。注意,默认情况下,Hive的数据库保存到hdfs://xxx:8020/user/hive/warehouse的目录下。

```
Hive>create database one;
```

查看Hive数据库:

```
Hive>show databases;
Default
One
```

查看HDFS目录:

```
$ hdfs dfs -ls /user/hive/warehouse
one.db
```

(2)使用MySQL数据库保存元数据

使用MySQL保存元数据的优点就是多个Hive Cli客户端可以同时登录,并操作相同的数据。这里,使用MySQL数据库来保存元数据,需要将mysql-connector.jar的驱动文件放到Hive3的lib目录下,然后配置hive-site.xml文件。

配置HIVE_HOME/config/hive-site.xml文件,将hive-site.template.xml文件重命名为hive-site.xml文件,然后输入以下内容:

```xml
<?xml version="1.0" encoding="UTF-8" standalone="no"?>
<?xml-stylesheet type="text/xsl" href="configuration.xsl"?>
<configuration>
    <property>
        <name>javax.jdo.option.ConnectionDriverName</name>
        <value>com.mysql.cj.jdbc.Driver</value>
    </property>
    <property>
        <name>javax.jdo.option.ConnectionURL</name>
        <value>jdbc:mysql://127.0.0.1:3306/hive?characterEncoding=UTF-8&amp;
serverTimezone=Asia/Shanghai&amp;useSSL=false
        </value>
    </property>
    <property>
        <name>javax.jdo.option.ConnectionUserName</name>
        <value>root</value>
    </property>
    <property>
        <name>javax.jdo.option.ConnectionPassword</name>
```

```xml
        <value>123456</value>
    </property>
</configuration>
```

与使用Derby数据库一样,首先需要初始化数据库操作,所以先执行初始化数据库命令:

```
$ /app/hive-3.1.2/bin/schematool -dbType mysql -initSchema
Initialization script completed
schemaTool completed
```

在初始化完成以后,会在MySQL的Hive数据库中看到已经初始化的数据表:

```
mysql> show tables;
+-------------------------------+
| Tables_in_hive                |
+-------------------------------+
| AUX_TABLE                     |
| BUCKETING_COLS                |
| CDS                           |
| COLUMNS_V2                    |
| COMPACTION_QUEUE              |
| COMPLETED_COMPACTIONS         |
| COMPLETED_TXN_COMPONENTS      |
| CTLGS                         |
| DATABASE_PARAMS               |
| DBS                           |
| DB_PRIVS                      |
...
```

创建完成以后,就可以使用hive命令,登录Hive的CLI操作数据了:

```
[hadoop@server201 app]$ hive
hive> show databases;
OK
Default
```

查看MySQL数据库中的表如下所示:

```
hive> show databases;
OK
default
one
```

步骤04 配置环境变量(可选)。

建议将Hive的bin目录配置到环境变量中,以便于更加方便地使用hive命令:

```
export HIVE_HOME=/app/hive-3.1.2
export PATH=$PATH:$HIVE_HOME/bin
```

现在就可以不用进入Hive的安装目录去登录Hive的客户端了。可以在任意目录下,执行hive命令即可以登录Hive的命令行模式:

```
$ hive
```

在登录hive命令行以后,执行类似于MySQL的命令show databases,即可显示当前所有的数据库:

```
hive> show databases;
OK
```

```
default
Time taken: 1.391 seconds, Fetched: 1 row(s)
```

上面命令输出的结果显示存在一个默认的数据库。还可以执行 show tables显示默认数据库下的所有表：

```
hive> show tables;
```

在启动Hive以后，会在Hadoop HDFS上出现/user/hive/warehouse的目录，这就是用于保存Hive数据的目录。到此为止，你的Hive已经可以运行了。

步骤 05 查看Hive命令的帮助。

同样地，通过--help参数可以查看所有Hive的帮助信息。

首先输入hive –help命令：

```
Usage ./hive <parameters> --service serviceName <service parameters>
```

其中，parameters可以为：

```
--auxpath : Auxiliary jars
--config : Hive configuration directory
--service : Starts specific service/component. cli is default
```

而ServiceName默认为cli。

如果需要显示版本，可以在--service后面添加version：

```
$ hive --service version
$ Hive-3.1.2
```

更多的帮助，请自行查看命令行说明。

6.2　Hive 的命令

Hive的命令很多类似SQL命令，但有些命令与SQL存在一些差异。Hive拥有自己的SQL语法。

1．创建一个数据库

Hive已经有了一个默认的数据库叫defalut，现在我们可以创建一个自己的数据库：

```
hive> create database one;
```

在创建完成这个数据库以后，就可以使用show databases显示Hive下的所有数据库：

```
hive> show databases;
OK
default
one
```

2．创建一个表

现在可以使用use one;命令进入one数据库，并在这个数据库下创建一个表：

```
hive> use one;
hive> create table stud(id int,name string);
```

在创建表以后，查看HDFS目录，会发现目录/user/hive/warehouse/one.db/stud，这就是刚才创建的表。

默认情况下，创建的表只会在元数据库（即MySQL或Derby）中保存一个数据结构，向里面写入数据时，默认会使用"^A"作为列的分隔符。建议使用row delimited来指定分隔符号。

3. 显示表结构

```
Hive> desc stud;
Hive> desc formatted stud;
```

你还可以使用show create table表名来显示表的结构：

```
01 hive> show create table stud;
02 OK
03 CREATE TABLE `stud`(
04     `id` int,
05     `name` varchar(30)
06 )  ROW FORMAT SERDE
07    'org.apache.hadoop.hive.serde2.lazy.LazySimpleSerDe'
08  STORED AS INPUTFORMAT
09    'org.apache.hadoop.mapred.TextInputFormat'
10  OUTPUTFORMAT
11    'org.apache.hadoop.hive.ql.io.HiveIgnoreKeyTextOutputFormat'
12  LOCATION
13    'hdfs://server201:8020/user/hive/warehouse/one.db/stud'
14  TBLPROPERTIES (
15    'transient_lastDdlTime'='1501079369')
16  Time taken: 0.886 seconds, Fetched: 13 row(s)
```

上面显示的表结构中，第09行用于声明这个文件的读取类。第10~11行声明输出数据类。第12~13行操作的目录。第14~15行为这个表最后修改的时间，当前显示为创建的时间。

甚至可以向这个表中保存数据（不建议这么做），由于Hive中的表对应的是文件，所以向Hive表中保存数据，就是向HDFS文件中保存数据。

向表中写入数据，虽然Hive SQL支持insert语句，但在真实的生产环境下，数据一般是从外部文件直接导入或是关联到Hive仓库的。这里执行insert语句只是为了做一个测试：

```
hive> insert into stud values(1,'SomeValue');
```

当执行insert语句向表中写入数据时，默认会执行Mapper任务向HDFS文件中写入数据。所以，执行上面的语句，会发现其实是执行了一个Mapper任务。再次声明，一般情况下我们不会执行insert写入数据，而是从外部（非HDFS文件）或是内部（HDFS文件）中导入数据。

现在可以执行select查询表中的数据：

```
hive> select * from stud;
OK
1    SomeValue
```

由此可以发现，里面已经存在了一行数据。再查询HDFS文件系统上的数据：

```
hive> dfs -cat /user/hive/warehouse/opt.db/stud/*;
1SomeValue
```

在Hive客户端命令行上,可以直接执行dfs命令,类似于执行hdfs dfs命令,只是省去了hdfs字样。通过上面的执行结果可以看出,里面已经写入了1SomeValue一行数据,且并没有分隔符号。现在我们将里面的文件000000_0下载到本地。

```
hive> dfs -get /user/hive/warehouse/opt.db/stud/* /home/hadoop/;
```

现在可以通过vim查看下载的文件:

```
$vim 000000_0
1^ASomeValue
```

由此可见,1与Jack之间是通过"^A"作为分隔符号的。现在我们可以在创建一个表时,指定数据之间的分隔符号。

创建一个表,并指定分隔符号:

```
hive> create table person(
    >    id int,
    >    name varchar(30)
    > )
    > row format delimited fields terminated by '\t';
```

注意上面语句中最后的分号";",如果没有分号,就像SQL语句没有结束一样,所以在Hive中分号也是语句结束的标记。上例中,设置"\t"(制表)符号为字段数据之间分隔的标记。

现在可以再写入一行数据,测试一下数据在HDFS文件中的分隔符号:

```
hive> insert into person(id,name) values(100,'Mary');
```

查看数据的分隔符号:

```
hive> dfs -cat /user/hive/warehouse/opt.db/person/*;
100    Mary
```

通过上面的结果,可以看出字段之间已经通过"\t"进行了分隔。

4. load data上传本地文件

load data命令用于上传一个本地文件到Hive的一个表中。其中local参数,用于加载本地磁盘上的一个文件。如果没有local参数,则会加载HDFS文件上的文件到Hive的表中。

默认情况下,程序处理的流程是先使用MapReduce将原始数据处理成一定格式的文本文件,然后再通过load data将数据加载到Hive的表中,最后使用HQL进行数据分析。

比如要将一个文件中的数据导入到上述的person表中,由于person表中字段之间的数据是用"\t"分隔的。因此,可以先通过vim创建一个数据文件,并用"\t"分隔里面的数据。

```
$ vim person.txt
101    Jack
102    Mary
103    Mark
104    Alex
```

现在使用load data命令,将数据导入到person表中:

```
hive> load data local inpath '/home/hadoop/person.txt' into table person;
```

在执行上面的导入语句以后,会在HDFS上发现person.txt这个文件,这个文件所在的目录为/user/hive/warehouse/opt.db/person/person.txt。

现在查询里面的数据:

```
hive> select * from person;
OK
100        Mary
101        Jack
102        Mary
103        Mark
104        Alex
```

也可以上传一个HDFS文件到person表中,使用load data命令,不添加local,即可从HDFS上加载数据。比如先将某个文件上传到HDFS:

```
hive> dfs -put /home/hadoop/person.txt /person.txt;
```

再使用load data将HDFS上的文件载入到Hive的表中。需要注意的是,文件上传完成以后,HDFS上的文件会被移动到Hive中,即会删除原目录下的文件。

```
hive> load data inpath '/person.txt' into table person;
```

5. 执行MapReduce任务

对数据进行统计count,将会执行一个MapReduce任务。代码如下:

```
hive> select count(*) from person;
OK
15
```

执行上面的查询语句,你会发现一个完整的MapReduce过程,并且最终直接将结果输出到控制台。

也可以将计算的结果输出到本地文件中去。注意,下例输出的count是一个目录,里面的文件才是输出的结果数据。

```
hive> insert overwrite local directory "/home/hadoop/count"
    > select count(*) from person;
```

也可指定导出的数据的分隔符号:

```
Hive>insert overwrite local directory '/home/hadoop/out001'
>row format delimited
>fields terminated by '\t'
select id,name from t1;
```

也可以将计算的结果输出到HDFS上,不用local参数即可。同样,/count是一个目录,里面文件中的数据,才是我们需要的结果:

```
hive> insert overwrite directory '/count' select count(1) from person;
```

执行一个过滤排序的查询,会执行MapReduce任务:

```
hive> select * from person where id>102 order by id;
Total MapReduce CPU Time Spent: 3 seconds 440 msec
OK
```

```
104        Alex
104        Alex
104        Alex
105        Mark
105        Mark
105        Mark
```

到此为止，你已经执行了Hive的一些命令。除"select *"不会生成MapReduce之外，其他的命令都会生成MapReduce任务。可见，Hive极大地简化了MapReduce的开发。

6.3 Hive 内部表

默认情况下，使用create table创建的表都是Hive的内部表。创建的内部表，均保存在/user/hive/warehouse目录下。执行drop table删除内部表时，会将数据及数据文件全部删除。

在元数据管理中，内部表在tbls元数据表中显示为MANAGED_TABLE，上一节所创建的表都是内部表，下面我们来看内部表的定义。

在Hive中，表都在HDFS的相应目录中存储数据。目录的名称是在创建表时自动创建并以表名来命名的，表中的数据都保存在该目录中。而且，数据以文件的形式存储在HDFS中。表的元数据会存储在数据库中，如Derby数据库或MySQL数据库。

创建表的语法：

```
CREATE [EXTERNAL] TABLE [IF NOT EXISTS] table_name
(col_name data_type [COMMENT col_comment], ...)
[COMMENT table_comment]
[PARTITIONED BY (col_name data_type [COMMENT col_comment], ...)]
[CLUSTERED BY (col_name, col_name, ...) INTO num_buckets BUCKETS]
[SORTED BY (col_name [ASC|DESC], ...)]
[ROW FORMAT DELIMITED row_format]
[STORED AS file_format]
[LOCATION hdfs_path];
```

语法说明如下：

① CREATE TABLE，创建一个名字为table_name的表。如果该表已经存在，则抛出异常；可以用IF NOT EXISTS关键字选项来忽略异常。

② 使用EXTERNAL关键字可以创建一个外部表，在建表的同时指定实际表数据的存储路径（LOCATION）。创建Hive内部表时，会将数据移动到数据仓库指定的路径；若创建Hive外部表，仅记录数据所在的路径，不对数据的位置做任何改变。在删除内部表时，内部表的元数据和数据会被一起删除；在删除外部表时，只删除外部表的元数据，但不删除数据。

③ (col_name data_type, ...)，创建表时要确定字段名及其数据类型，数据类型可以是基本数据类型，也可以是复杂数据类型。COMMENT为表和字段添加注释描述信息。

④ PARTITIONED BY，创建分区表。

⑤ CLUSTERED BY，创建桶表。

⑥ SORTED BY，排序。

⑦ ROW FORMAT DELIMITED，用于指定表中数据行和列的分隔符及复杂数据类型数据的分隔符。这些分隔符必须与表数据中的分隔符完全一致。

- [Fields Terminated By Char]，用于指定字段分隔符。
- [Collection Items Terminated By Char]，用于指定复杂数据类型Map、Struct和Array的数据分隔符。
- [Map Keys Terminated By Char]，用于指定Map中的Key与Value的分隔符。
- [Lines Terminated By Char]，用于指定行分隔符。

⑧ STORED AS，指定表文件的存储格式，如TextFile格式、SequenceFile格式、ORC格式和Parquet格式等。如果文件数据是纯文本的，可以使用TextFile格式，这种格式是默认的表文件存储格式。如果数据需要压缩，可以使用SequenceFile格式等。

⑨ LOCATION，用于指定所创建表的数据在HDFS中的存储位置。

不带EXTERNAL关键字创建的表是管理表，有时也称为内部表。Hive表是归属于某个数据仓库的，默认情况下Hive会将表存储在默认数据仓库中，也可以使用Use命令切换数据仓库，将所创建的表存储在切换后的数据仓库中。

使用drop table删除内部表时，表的元数据和表数据文件同时被删除，读者可以自行测试一下。

下面是内部表创建举例。

1．示例说明

创建内部表test，并将本地/opt/datas/test.txt目录下的数据导入Hive的test(id int, name string)表中。

在/opt/datas目录下准备数据，创建test.txt文件并添加数据：

```
hadoop@SYNU:/opt/datas$ vim test.txt
101 Bill
102 Dennis
103 Doug
104 Linus
105 James
106 Steve
107 Paul
108 Ford
```

test.txt文件中的数据以Tab键分隔。

2．Hive实例操作

（1）启动Hive

```
hadoop@SYNU:/usr/local/hive$ bin/hive
```

（2）显示数据仓库

```
hive(default)>show databases;
```

（3）切换到hivedwh数据仓库

```
hive(default)>use hivedwh;
```

（4）显示hivedwh数据仓库中的表

```
hive(hivedwh)>show tables;
```

（5）创建test表，并声明文件中数据的分隔符

```
hive(hivedwh)>create table test(id int, name string)
row format delimited fields terminated by '\t';
```

（6）加载/opt/datas/test.txt 文件到test表中

```
hive(hivedwh)>load data local inpath '/opt/datas/test.txt' into table test;
```

（7）Hive查询结果

```
hive(hivedwh)>select id,name from test;
OK
```

6.4 Hive 外部表

如果数据已经存在于HDFS上，则可以通过创建外部表的方式与HDFS上的数据建立关系。默认通过create table创建表为内部表（managed_table）。在元数据tbls表中有一列TBL_TYPE，如果其值为MANAGED_TABLE，则为内部表。

通过create external table可以创建一个外部表。在创建外部表时，需要指定HDFS上的一个目录，创建外部表与目录关系。

```
hive> create external table ext_person(id bigint,name string)
    > row format
    > delimited fields terminated by '\t'
    > location '/test/person';
OK
Time taken: 0.149 seconds
```

在上面的代码中，name可以使用varchar(N)，也可以使用string。注意，最后location '/test/person'用于指定HDFS的目录，在此目录下包含多个txt类型的文件，里面保存了id和name以\t分隔的数据。现在查看tbls表的tbl_type列，值为EXTERNAL_TABLE，即为外部表。

查询外部表就像是查询内部表一样，可以获取数据结果：

```
hive> select * from ext_person;
OK
101     Jack
102     Mary
103     Mark
104     Alex
```

外部表创建之后，tbls元数据表中保存外部表的信息：table_type=EXTERNAL_TABLE。

使用drop table删除表之后，外部表所引用的HDFS目录不会被删除，当然，HDFS目录下的相关数据也不会被删除。

6.5　Hive 表分区

Hive表分区的背景如下：

（1）在Hive Select查询中一般会扫描整个表内容，会消耗很多时间做没必要的工作。有时候只需要扫描表中关心的一部分数据，因此建表时引入了Partition，即分区的概念。

（2）分区表指的是在创建表时指定了Partition分区空间的表。

（3）如果需要创建有分区的表，需要在create表的时候传入可选参数partitioned by关键字。

6.5.1　分区技术细节

（1）一个表可以拥有一个或者多个分区，每个分区以文件夹的形式出现在HDFS文件系统的目录中。

（2）表和列名不区分大小写。

（3）分区是以字段的形式在表结构中存在，通过describe table命令可以查看到字段的存在，但是该字段不存放实际的数据内容，仅仅是分区的表示。

（4）建表的语法（建分区可参见PARTITIONED BY参数）为：

```
CREATE [EXTERNAL] TABLE [IF NOT EXISTS] table_name [(col_name data_type [COMMENT col_comment], ...)] [COMMENT table_comment] [PARTITIONED BY (col_name data_type [COMMENT col_comment], ...)] [CLUSTERED BY (col_name, col_name, ...) [SORTED BY (col_name [ASC|DESC], ...)] INTO num_buckets BUCKETS] [ROW FORMAT row_format] [STORED AS file_format] [LOCATION hdfs_path]
```

（5）分区建表分为2种：一种是单分区，也就是说在表文件夹目录下只有一级文件夹目录；另外一种是多分区，表文件夹下出现多文件夹嵌套模式。

① 单分区建表语句中使用partitioned by即可。比如以下语句通过partitioned by创建了一个分区grade（年级），将会在/user/hive/warehouse/stud目录下创建多个不同的分区目录，如01年级的数据就会保存到/user/hive/warehouse/stud/01下。

```
hive> create table stud(id int,name string) Partitioned by (grade string)
    > row format delimited fields terminated by '\t';
```

现在我们可以使用insert插入一些数据，虽然不建议这么做，但这样做可以快速了解分区的存储形式：

```
hive> insert into stud(id,name,grade) values(2,'Mike','2002');
hive> insert into stud(id,name,grade) values(1,'Jack','2001');
```

以上两个语句会开启MR程序，写入两行数据。插入数据以后，查看HDFS的目录结构，即可以看到 2001和2002两个子目录。为了显示表头，可以设置让Hive显示表头。

```
hive> set hive.cli.print.header=true;
```

查询数据：

```
hive> select * from stud;
OK
stud.id    stud.name    stud.grade
1    Jack    2001
2    Mike    2002
```

查看HDFS上的数据，可见分区最后形成了目录：

```
/user/hive/warehouse/stud/grade=2001
/user/hive/warehouse/stud/grade=2002
```

导入数据时指定分区值。在导入数据时，只要在最后添加Partition(..)，即可以将导入的数据添加到指定的分区。

```
hive> load data local inpath '/root/stud.txt' into table stud partition (grade='2008');
```

② 创建多个分区：

以下示例表创建三个列，id、name、major（专业）以及grade四个列。其中major、grade为分区数据，即根据专业、年级对学生信息进行分区。

```
hive> create table students(id int,name string)
    > partitioned by (major string,grade string)
    > row format delimited fields terminated by '\t';
```

上面语句创建两个分区，一个为major即按专业分区，另一个grade再按年级进行分区。

对于这个双分区表，表结构中新增加了major和grade两列。此时，保存到HDFS上的文件系统显示如下所示。注意，只有保存了数据以后，才会存在以下的结构，且会根据保存数据的不同，分区数据会显示为不同的目录。

```
/user/hive/warehouse/opt.db/students/major=Java/grade=2017
```

（6）添加分区是指在创建表时没有创建分区列，通过修改表的方式，添加新的分区信息。

```
ALTER TABLE table_name ADD partition_spec [ LOCATION 'location1' ] partition_spec
[ LOCATION 'location2' ] ... partition_spec: : PARTITION (partition_col =
partition_col_value, partition_col = partition_col_value, ...)
```

用户可以用 ALTER TABLE ADD PARTITION 来向一个表中增加分区。当分区名是字符串时，需要加单引号。比如：

```
hive> alter table student add partition(major='Java');
```

（7）删除分区数据：

```
ALTER TABLE table_name DROP partition_spec, partition_spec,...
```

用户可以用 ALTER TABLE DROP PARTITION 来删除分区及里面的数据。分区的元数据和数据将被一并被删除。比如：

```
hive> alter table student drop partition(major="Java");
```

（8）数据加载进分区表。

将已有数据保存到指定分区中，它的语法是：

```
LOAD DATA [LOCAL] INPATH 'filepath' [OVERWRITE] INTO TABLE tablename [PARTITION
```

```
(partcol1=val1, partcol2=val2 ...)]
```

现在加载数据到指定的分区中：

```
hive> load data local inpath '/home/hadoop/person.txt' into table students
    > partition(major='Java',grade='2017');
```

当数据被加载至表中时，不会对数据进行任何转换。load操作只是将数据复制至Hive表对应的位置。数据加载时，在表下自动创建一个目录，文件存放在该分区下。

即当查询保存到分区中的数据时，数据的内容与加载之前的数据完全一样，比如：

```
hive> dfs -cat \
 /user/hive/warehouse/opt.db/students/major=Java/grade=2017/*;
    101     Jack
    ...
```

但当执行查询时，会将分区的数据当作表的一部分查询出来，比如：

```
hive> select * from students;
OK
101     Jack    Java    2017
102     Mary    Java    2017
103     Mark    Java    2017
104     Alex    Java    2017
```

（9）基于分区的查询语句。

现在你可以使用load data导入更多的数据，其后在查询语句中使用分区的字段作为查询条件。

```
hive> select * from students where major='Oracle';
OK
301     Jack    Oracle    2017
302     Mary    Oracle    2017
303     Alex    Oracle    2017
304     Smith   Oracle    2017
```

（10）查看分区语句。

可以通过Show Partitions查看一个表的分区信息：

```
hive> show partitions students;
OK
major=Java/grade=2017
major=Oracle/grade=2017
Time taken: 0.235 seconds, Fetched: 2 row(s)
```

6.5.2 分区示例

（1）在Hive中，表中的一个Partition对应于表下的一个目录，所有的Partition的数据都存储在最子集的目录中。

（2）总的来说，Partition就是辅助查询，缩小查询范围，加快数据的检索速度，以及对数据按照一定的规格和条件进行管理。

以下示例创建一个班级表，其拥有三个字段，id、name和classname（班级名称），以classname作为分区：

```
hive> create table students(id bigint,name string)
    > partitioned by (classname string)
    > row format delimited fields terminated by '\t';
```

在创建成功以后,数据保存到HDFS的/user/hive/warehouse/students目录下。

现在把数据保存到students表中,先使用vi编辑一个文本文件,里面的内容如下:

```
vi studs.txt
101        Jack
102        Mary
...
```

然后使用load data将数据保存到students表中,由于students表是有分区的,因此在load data时必须指定分区信息,注意最后的partition(classname="Java")。

```
hive> load data local inpath '/home/wangjian/studs.txt'
    > into table students partition(classname="Java");
```

可以将/home/hadoop路径下的学生信息进行加载,并指定不同的分区信息:

```
hive> load data local inpath '/home/hadoop/studs.txt'
    >into table students partition(classname="oracle");
```

在load数据时,可以通过overwite关键字重新设置某个分区下的所有数据:

```
hive> load data local inpath '/home/wangjian/studs2.txt'
    >overwrite into table students partition(classname="Oracle");
```

现在查看HDFS上的目录结构如下:

```
hive> dfs -ls /user/hive/warehouse/students;
Found 2 items
/user/hive/warehouse/opt.db/students/classname=Java
/user/hive/warehouse/opt.db/students/classname=Oracle
```

可见,在students目录里面出现了两个子目录,分别为classname=Java和classname=Oracle,即为两个分区目录。

现在查询一下所有的数据:

```
hive> select * from students;
101     Jack    Java
102     Mary    Java
301     Jack    Oracle
302     Mary    Oracle
303     Alex    Oracle
```

显示Partition的信息:

```
hive> show partitions students;
classname=Java
classname=Oracle
```

创建具有多个分区的表,如某个班级是哪一年级的哪一个专业,以下即指定两个分区信息:

```
hive> create table classes(id bigint,name string)
    > partitioned by(grade string,major string)
    > row format delimited fields terminated by '\t';
```

导入数据，请先使用vim编辑一个文件如cls1.txt，加入一些数据，然后再做导入：

```
hive> load data local inpath '/home/hadoop/cls1.txt'
>overwite into table classes partition(grade='2016',major='computer');
```

导入数据以后，会出现两层目录结构：

```
/user/hive/warehouse/classes/grade=2016/major=computer
```

导入更多数据，并做一个查询：

```
hive> load data local inpath '/home/hadoop/cls2.txt'
>overwrite into table classes partition(grade='2017',major='ui');
hive> select * from classes;
1    Java1    2016    computer
2    Java2    2016    computer
9    UI1      2017    ui
10   UI2      2017    ui
```

添加一个新的分区，如添加一个学院信息，即哪一个班级、属于哪一个学院：

```
hive> alter table classes add partition \
 (grade='2015',major='computer');
```

添加一个新的分区是指在当前表的目录下创建分区目录的过程。它本质上不是修改表结构，而是添加数据目录的过程。

在语句执行完成以后，新的HDFS目录结构为：

```
/user/hive/warehouse/classes/grade=2015/major=computer
```

删除一个分区：

```
hive> alter table classes drop \
 partition(grade='2015',major='computer');
```

6.6 查询示例汇总

创建一个表，并指定列分割方式：

```
hive> create table person(id string,name string,age int,sex string)
    > row format delimited
    > fields terminated by '\t';
```

创建一个表，并指定存储格式为textfile：

```
hive> create table car(id string,name string,price double,pid string)
    > row format delimited
    > fields terminated by '\t'
    > stored as textfile;
```

关联查询：

```
hive> select p.id,p.name,c.id,c.name from person p inner join car c on p.id=c.pid;
```

使用group by：

```
hive> select count(1),pid from car group by pid;
```

使用having：

```
hive> select count(1),pid from car group by pid having count(1)>=2;
```

子查询：

```
hive> select * from person where id in(select pid from car group by pid having count(1)>=2);
```

相关子查询：

```
hive> select * from person where (select count(1) from car where person.id=car.pid)=2;
```

6.7 Hive 函数

Hive内部定义了很多的函数，这些函数都是通过Hive的FunctionRegistry类注册的。在IDE中查看此类的源代码，需要添加Hive Query Language的依赖：

```
<dependency>
    <groupId>org.apache.hive</groupId>
    <artifactId>hive-exec</artifactId>
    <version>3.1.2</version>
</dependency>
```

添加依赖之后，就可以查看FunctionRegistry这个类的源代码了，你将会发现大量类似于以下的函数注册代码：

```
// registry for system functions
private static final Registry system = new Registry(true);
static {
    system.registerGenericUDF("concat", GenericUDFConcat.class);
    system.registerUDF("substr", UDFSubstr.class, false);
    system.registerUDF("substring", UDFSubstr.class, false);
    system.registerGenericUDF("substring_index", GenericUDFSubstringIndex.class);
    system.registerUDF("space", UDFSpace.class, false);
    system.registerUDF("repeat", UDFRepeat.class, false);
    system.registerUDF("ascii", UDFAscii.class, false);
    system.registerGenericUDF("lpad", GenericUDFLpad.class);
    system.registerGenericUDF("rpad", GenericUDFRpad.class);
    system.registerGenericUDF("levenshtein", GenericUDFLevenshtein.class);
    system.registerGenericUDF("soundex", GenericUDFSoundex.class);
    ...
```

上面的代码就是向Hive系统注册函数。在执行HQL语句的过程中，可以使用Hive已经定义的函数。我们可以使用show functions语句查看Hive中所有的内部函数：

```
hive> show functions;
```

```
...
<
<=
<=>
..
abs
acos
add_months
aes_decrypt
aes_encrypt
...
```

通过show functions可以列出Hive内部已经注册的所有函数，可以看到运算符号都已经被注册成了Hive的函数，包括abs、acos等数学函数和其他更多的函数。

也可以使用"describe function 函数名称;"来查看具体某一个函数的用法。比如我们先来查看一下运算符号"="的具体说明：

```
hive> describe function =;
OK
a = b - Returns TRUE if a equals b and false otherwise
Time taken: 0.012 seconds, Fetched: 1 row(s)
```

函数uuid的功能：

```
hive> describe function uuid;
OK
uuid() - Returns a universally unique identifier (UUID) string.
Time taken: 0.015 seconds, Fetched: 1 row(s)
```

以下是Hive中的函数及其示例。

1. 关系运算符号

在FunctionRegistry类中，可以找到注册关系运算位的源代码：

```
HIVE_OPERATORS.addAll(Arrays.asList(
    "+", "-", "*", "/", "%", "div", "&", "|", "^", "~",
    "and", "or", "not", "!",
    "=", "==", "<=>", "!=", "<>", "<", "<=", ">", ">="));
```

通过HIVE_OPERATORS.addAll添加的都是操作符号。其中"+"、"-"、"*"、"/"、"%"、"div"、"&"、"|"、"^"、"~"等符号大多用于select子句中。

除"/"运算符号运算的结果为double类型：

```
hive> select 10/3;
3.3333333333333335
```

div操作符号也是除操作，运算的结果为整数类型：

```
hive> select 10 div 3;
3
```

%为取模操作，即计算余数：

```
hive> select 10 % 3;
1
```

与（&）操作，二进制运算，两个值必须都为1时结果才是1，否则为0。下面示例中，2的二进制为10，1二进制为01，则10 & 01结果为00，即为0。

```
hive> select 2 & 1;
0
```

或（|）操作，二进制运算，两个值只要有一个为1，即为1。下面示例中，2的二进制为10，1的二进制为01，则 10|01的结果为11，即结果为3。

```
hive> select 2 | 1;
3
```

异或（^）运算，二进制运算，只要两个值不一样，则为1。两个值一样时为0。下面示例中，10^01的结果为11即结果为3。

```
hive> select 2^1;
3
```

按位取反（~）操作符号，二进制运算，~1为0，~0为1。下面示例中，2的二进制为10，所以~10的值为11111111111111111111111111111101，即结果为-3。

```
hive> select ~2;
-3
```

操作符号"="、"=="、"<=>"、"!="、"<>"、"<"、"<="、">"、">="用在where子句中，均用于比较。相等比较"="、"=="、"<=>"，具有相同的功能。

```
hive> select * from students where id==301;
hive> select * from students where id=="301";
hive> select * from students where id<=>"301";
```

操作符号"and"、"or"、"not"、"!"，用于where子句中，分别进行与、或、非运算：

```
hive> select * from students where id=301 and major='Oracle';
hive> select * from students where id=301 or major='Java';
```

2．更多函数

Hive拥有丰富的内置函数。由于函数太多，以下仅为读者展示一部分。

（1）array

根据给定的元素，创建一个数组对象。

语法：array(n0, n1,...)

以下创建一个字符串数组对象：

```
hive> select array('Jack','Mary');
["Jack","Mary"]
```

以下创建一个整数的数组对象：

```
hive> select array(1,2,3);
[1,2,3]
```

以下是由于包含一个字符串，所以创建一个字符串数组对象：

```
hive> select array(1,'Jack');
["1","Jack"]
```

(2) array_contains

判断给定的元素是否在数组中存在。

语法：array_contains(array, value)

```
hive> select array_contains(array('Jack','Mary'),'Mary');
true
```

(3) bin

将一个bigint类型的数转成二进制。

语法：bin(n) - returns n in binary

```
hive> select bin(2);
10
```

(4) ceil

返回大于当前数的最小整数。

语法：ceil(x)

```
hive> select ceil(2.3);
3
```

(5) current_date

返回当前时间。

语法：current_date()

```
hive> select current_date();
2021-3-18
```

(6) current_timestamp

返回当前的时间戳。

语法：current_timestamp()

```
hive> select current_timestamp();
2021-3-18 21:37:44.271
```

(7) explode

将数组元素转换成多行显示。

语法：explode(a)

```
hive> select  explode(array("Jack","Mary"));
Jack
Mary
```

也可以将一个map转换成多行显示。

```
hive> select  explode(map("name","Jack","age",34));
name    Jack
age     34
```

(8) get_json_object

根据指定的路径，解析出json字符串中的对象，json中必须是" "（双引号，即标准的json串），path部分必须是$.开始。如果只输入$表示当前整个json对象。

语法：get_json_object(json_txt, path)

```
hive> select get_json_object('{"name":"Jerry"}','$.name');
OK
Jerry
hive> select get_json_object('["Jack","Rose"]','$[1]');
OK
Rose
```

(9) map

创建一个map对象。

语法：map(key0, value0, key1, value1, ...)

```
hive> select map("name","Jack","age",34);
{"name":"Jack","age":"34"}
```

(10) split

根据指定的正则表达式将字符串转换成数组，正则表达式中要使用\\（两个斜线）。

语法：split(str, regex)

```
hive> select split('Jack Mary Rose','\\s+');
["Jack","Mary","Rose"]
hive> select explode(split('Jack Mary Rose','\\s+'));
Jack
Mary
Rose
```

3. 使用Hive函数实现WordCount

首先使用vim创建一个文本文件：

```
$vim /hone/hadoop/notes.txt
```

里面的内容如下：

```
Hello this is a text for something
to tell you about how to process
wordcount in hive.
And you must be import into this file
into hive.
```

现在创建一个Hive表，只包含一个列，且分隔符号为'\r\n'，即回车换行。

```
hive>create table notes(text string)
>row format
>delimited fields terminated by '\r\n';
```

将notes.txt文件导入到notes表中：

```
hive> load data local inpath '/home/hadoop/notes.txt' into table notes;
```

测试查询是否是5行数据：

```
hive> select * from notes;
Hello this is a text for something
to tell you about how to process
wordcount in hive.
And you must be import into this file
into hive.
Time taken: 0.159 seconds,
Fetched: 5 row(s)
```

再创建一个表，用于保存每一个单词：

```
hive> create table word(w string)
    > row format
    > delimited fields terminated by '\r\n';
```

现在我们需要将notes表中的每一行数据，按空格进行split，然后再转换成行，保存到word表中去。以下示例使用insert overwrite语句，会先将word表中的数据删除，然后再写入新的数据；如果使用insert into，将会是追加数据。此语句会引发一个MapReduce计算。

```
hive> insert overwrite table word select explode(split(text,'\\s+')) from notes;
Total MapReduce CPU Time Spent: 2 seconds 740 msec
OK
Time taken: 22.725 seconds
```

先对word表进行查询，发现已经将单词都保存到了word表中：

```
hive> select * from word;
OK
this
is
a
text
for
something
to
tell
you
...
```

现在对word表进行count查询：

```
hive> select count(w),w from word group by w;
```

这个查询将会启动MapReduce，并最终输出以下结果（部分内容略去）。

```
1    And
1    Hello
...
2    you
```

也可以将计算的结果保存到指定目录下：

```
hive> insert overwrite directory '/out001/' select concat(w,'\t',count(w)) from word group by w ;
```

直接在Hive里面查询数据：

```
hive> dfs -cat /out001/*;
And    1
```

```
a       1
about   1
be      1
file    1
for     1
hive.   2
...
```

6.8　Hive 自定义函数

UDF（User Defined Function）是Hive中的自定义函数。当Hive中自有函数不能达到我们的业务要求时，可以通过自定义UDF实现需要的业务逻辑。

Hive中已经存在很多的函数，这些函数都是Hive定义的，可以在Hive CLI命令行下查看。关于这一点，上一节已经说过，这里不再赘述。

```
SHOW FUNCTIONS;   显示所有函数
DESCRIBE FUNCTION <function_name>;   显示某个函数的说明
DESCRIBE FUNCTION EXTENDED <function_name>;   显示某个函数的说明，及示例程序
```

关于更多的Hive函数，可以通过以下官方网站查看：

https://cwiki.apache.org/confluence/display/Hive/LanguageManual+UDF

Hive中的函数可以分为UDF（即用户自定义函数）、UDAF（User-Defined Aggregation Function，即用户自定义的聚合函数，如sum、count等）、UDTF（User-Defined Table-Generating Functions，即用户自定义的表创建函数）。

Hive3的自定义函数比Hive1自定义函数相对来说更复杂。下面讲解一下如何在Hive3中创建一个自定义函数。

步骤01 创建Java项目添加依赖。

只需要添加hive-exec这一个依赖包即可：

```xml
<dependency>
    <groupId>org.apache.hive</groupId>
    <artifactId>hive-exec</artifactId>
    <version>3.1.2</version>
</dependency>
```

步骤02 开发Java类。

开发自定义函数，在Hive2以后，可以继承UDF，继承UDF的开发相对比较简单。而继承GenericUDF则相对比较复杂。虽然UDF类已经不再建议被使用，但是Hive自己的很多函数还是通过继承UDF来实现的。因此，开发UDF可以根据自己的情况来实现。以下是自定义add(x,y)函数的两种开发形式。

（1）通过继承UDF开发的自定义函数，如代码6.1所示。

代码6.1　Add.java

```
01 package org.hadoop.udf;
```

```
02  import org.apache.hadoop.hive.ql.exec.UDF;
03  public class Add extends UDF {
04      public Integer evaluate(Integer a, Integer b) {
05          System.out.println("a:" + a + ",b:" + b);
06          Integer c = a + b;
07          System.out.println("c is:" + c);
08          return c;
09      }
10  }
```

(2) 通过继承GenericUDF开发的自定义函数，如代码6.2所示。

代码6.2 AddFun.java

```
01  package org.hadoop.udf;
02  import org.apache.hadoop.hive.ql.exec.UDFArgumentException;
03  import org.apache.hadoop.hive.ql.metadata.HiveException;
04  import org.apache.hadoop.hive.ql.udf.generic.GenericUDF;
05  import org.apache.hadoop.hive.serde2.objectinspector.ObjectInspector;
06  import org.apache.hadoop.hive.serde2.objectinspector.ObjectInspectorConverters;
07  import org.apache.hadoop.hive.serde2.objectinspector.PrimitiveObjectInspector;
08  import org.apache.hadoop.hive.serde2.objectinspector.primitive.
    PrimitiveObjectInspectorFactory;
09  import org.apache.hadoop.io.IntWritable;
10  public class AddFun extends GenericUDF {
11      private ObjectInspectorConverters.Converter converter1;
12      private ObjectInspectorConverters.Converter converter2;
13      @Override
14      public ObjectInspector initialize(ObjectInspector[] arguments) throws
    UDFArgumentException {
15          if (arguments.length != 2) {
16              throw new UDFArgumentException("需要两个参数，但只有：" + arguments.length);
17          }
18          if (arguments[0].getCategory() != ObjectInspector.Category.PRIMITIVE) {// 判
    断是否是基本类型
19              throw new UDFArgumentException("第一个参数，不是基本类型，而是：" +
    arguments[0].getTypeName());
20          }
21          if (arguments[1].getCategory() != ObjectInspector.Category.PRIMITIVE) {
22              throw new UDFArgumentException("第二个参数，不是基本类型，而是：" +
    arguments[1].getTypeName());
23          }
24          PrimitiveObjectInspector o1 = (PrimitiveObjectInspector) arguments[0];
25
26          if (o1.getPrimitiveCategory() !=
    PrimitiveObjectInspector.PrimitiveCategory.INT) {// 如果不是INT类型
27              throw new UDFArgumentException("第一个参数，不是int类型，而是：" +
    o1.getTypeName());
28          }
29          o1 = (PrimitiveObjectInspector) arguments[1];
30          if (o1.getPrimitiveCategory() !=
    PrimitiveObjectInspector.PrimitiveCategory.INT) {
31              throw new UDFArgumentException("第二个参数不是int类型，而是：" +
    o1.getTypeName());
32          }
```

```
33              //声明转换类型为基本的int类型,每一个参数,必须声明一个 Convert否则转换出错
34              converter1 = ObjectInspectorConverters.getConverter(arguments[0],
35                  PrimitiveObjectInspectorFactory.writableIntObjectInspector);
36              converter2 = ObjectInspectorConverters.getConverter(arguments[1],
37                  PrimitiveObjectInspectorFactory.writableIntObjectInspector);
38              // 指定返回的类型
39              return PrimitiveObjectInspectorFactory.writableIntObjectInspector;
40          }
41          @Override
42          public Object evaluate(DeferredObject[] arguments) throws HiveException {
43              Object o1 = arguments[0].get();
44              Object o2 = arguments[1].get();
45              System.out.println("o1:"+o1+",o2:"+o2);
46              o1 = converter1.convert(o1);
47              o2 = converter2.convert(o2);
48              System.out.println("转换以后:"+o1+","+o2);
49              Integer a1 = ((IntWritable)o1).get();
50              Integer a2 = ((IntWritable)o2).get();
51              System.out.println("转成int以后为: "+a1+","+a2);
52              Integer a3= a1+a2;
53              System.out.println("结果: "+a3);
54              return new IntWritable(a3);
55          }
56          @Override
57          public String getDisplayString(String[] children) {
58              return "AddFun";
59          }
60      }
```

步骤03 将程序打成jar包。

将项目打成 jar包,如udf.jar,上传到服务器上去。使用Maven的打包命令,在项目的根目录下执行以下语句:

```
mvn package
[INFO] --- maven-jar-plugin:2.4:jar (default-jar) @ chapter08 ---
E:\gits\gitee\hadoop3.2\hadoop\chapter08\target\chapter06-1.0.jar
[INFO] BUILD SUCCESS
[INFO] Total time:  7.765 s
[INFO] Finished at: 2021-03-18T22:05:12+08:00
```

现在将chapter06-1.0.jar修改成udf.jar并上传到Linux服务器。

步骤04 添加jar。

使用add jar命令,类似于给Hive设置classpath的目录,这样方便让Hive可以找到UDF所在的jar包。但这种方式在下次重新启动Hive时将失效。

```
hive> add jar /home/hadoop/udf/udf.jar;
Added [/home/hadoop/udf/udf.jar] to class path
Added resources: [/home/hadoop/udf/udf.jar]
```

如果希望添加的jar长期有效,可以将jar放到HIVE_HOME/auxlib目录下。

也可以添加HDFS上的jar包:

```
hive> add jar hdfs://server201:8020/udf/udf.jar;
```

步骤05 添加函数并运行。

通过create temporary可以创建一个临时函数，函数名为add，此函数将使用类cn.hive.fun.AddFun。通过create temporary创建的函数，当退出Hive时，函数将失效。

```
hive> create temporary function add as "org.hadoop.udf.Add";
OK
Time taken: 0.02 seconds
```

如果希望自定义的函数长久有效，可以通过修改Hive的FunctionRegistry类实现。

现在就可以执行这个函数，并进行计算，由于我们在代码中写了system.out输出语句，所以会输出一些控制台信息，但只要计算结果正确，就说明UDF函数开发完成了。

```
hive> select add(2,3);
a:2,b:3
c is:5
a:2,b:3
c is:5
a:2,b:3
c is:5
OK
5
Time taken: 0.115 seconds, Fetched: 1 row(s)
```

使用同样的方法，可以运行AddFun.java中定义的函数。

```
hive> create temporary function AddFun as "org.hadoop.udf.AddFun";
OK
Time taken: 0.009 seconds
hive> select AddFun(2,4);
o1:2,o2:4
转换以后：2,4
转成int以后为：2,4
结果：6
o1:2,o2:4
转换以后:2,4
转成int以后为：2,4
结果：6
o1:2,o2:4
转换以后:2,4
转成int以后为：2,4
结果：6
OK
6
Time taken: 0.109 seconds, Fetched: 1 row(s)
```

注意：由于大多数Hive的查询，会将生成的结果保存到HDFS上。如以下命令，会将执行结果保存到HDFS上。

```
hive> insert overwrite directory '/hiveout1' select add(1,4);
```

所以，在开发UDF时，可以接收LongWriteable/Text等Hadoop可以处理的数据类型。笔者也建议数据类型使用Hadoop可以处理的类型。

6.9 Hive 视图

与传统数据库类似,Hive也可以创建视图(view)。

创建视图:

```
Hive> create view v1 as select * from person where sex='1';
```

查看视图:

```
Hive>show views;
```

查询视图:

```
Hive> select * from v1;
```

删除视图:

```
Hive>drop view v1;
```

6.10 hiveserver2

hiveserver2的特点是维护一个端口,并让用户维护一个长连接,就是像JDBC那样,并可以通过TCP在网络上传递数据。默认hiveserver2的端口为10000。

配置hiveserver2需要修改hive-site.xml文件,同时需要将Hadoop的core-site.xml和hdfs-site.xml文件放到hive-3.1.2/lib目录下。

步骤01 修改Hadoop的配置文件。

修改Hadoop的core-site.xml文件,添加以下内容。其中hadoop.proxyuser后面为用户名,这里将此值设置为hadoop。

```xml
<property>
    <name>hadoop.proxyuser.hadoop.hosts</name>
    <value>*</value>
</property>
<property>
    <name>hadoop.proxyuser.hadoop.groups</name>
    <value>*</value>
</property>
```

配置说明如下:

- hadoop.proxyuser.hadoop.hosts:指定hadoop用户可以在所有主机上访问。
- hadoop.proxyuser.hadoop.groups:指定hadoop用户不限制分组。

步骤02 复制文件。

由于hiveserver2在启动时需要读取Hadoop的配置信息,所以这里我们将core-site.xml文件和

hdfs-site.xml文件放到hive-3.1.2/conf目录下。直接使用Linux的ln创建一个文件的硬连接即可。

```
$ ln /app/hadoop-3.2.2/etc/hadoop/core-site.xml
/app/hive-3.1.2/conf/core-site.xml
$ ln /app/hadoop-3.2.2/etc/hadoop/hdfs-site.xml
/app/hive-3.1.2/conf/hdfs-site.xml
```

步骤03 配置hive-site.xml文件。

在原来已经配置了MySQL数据连接的基础上，再在hive-site.xml文件中添加以下配置：

```
<property>
    <name>hive.metastore.local</name>
    <value>false</value>
</property>
<property>
    <name>hive.metastore.uris</name>
    <value>thrift://server201:9083</value>
</property>
```

步骤04 启动hiveserver2。

首先必须启动的是metastore，metastore是Hive的存储服务，上面配置的使用MySQL保存元数据信息，可以理解为metastore的主要工作。

```
$ /app/hive-3.1.2/bin/hive --service metastore &
```

然后再启动hiveserver2，启动服务后将默认占用10000端口。

```
$/app/hive-3.1.2/bin/hive --service hiveserver2 &
```

查看hiveserver2的启动日志，默认放到/tmp/hadoop/hive.log里面。

```
2021-03-19 14:11:24: Starting HiveServer2
Hive Session ID = 4558e81a-0104-4f80-9f64-817e7a7f2244
```

步骤05 连接及操作。

启动hiveserver2之后，就可使用beeline连接Hive。通过beeline命令登录Hive，其中-u参数用于指定url，-n参数用于指定一个用户名称。登录成功后，将显示beeline的命令行：

```
[hadoop@server201 app]$ beeline -u jdbc:hive2://server201:10000 -n hadoop
Beeline version 3.1.2 by Apache Hive
0: jdbc:hive2://server201:10000>
```

登录成功以后，即可进行SQL操作：

```
0: jdbc:hive2://server201:10000> show databases;
+----------------+
| database_name  |
+----------------+
| default        |
| one            |
..
```

查询表中的数据：

```
0: jdbc:hive2://server201:10000> select * from stud;
| stud.id  | stud.name |
+----------+-----------+
```

```
| 1       | Jack      |
....
```

执行一个聚合操作，同样也会执行MapReduce操作：

```
0: jdbc:hive2://server201:10000> select count(1) from stud;
INFO  : Total jobs = 1
INFO  : Launching Job 1 out of 1
+------+
| _c0  |
+------+
| 7    |
+------+
1 row selected (20.608 seconds)
```

6.11 使用 JDBC 连接 hiveserver2

开启hiveserver2之后的好处是通过JDBC也可以连接Hive，就像是使用JDBC连接MySQL一样。

代码 6.3　HiveJdbc.java

```
01 package org.hadoop.hive;
02 import java.sql.Connection;
03 import java.sql.DriverManager;
04 import java.sql.ResultSet;
05 import java.sql.Statement;
06 public class HiveJdbc {
07     public static void main(String[] args) throws Exception {
08         Class.forName("org.apache.hive.jdbc.HiveDriver");
09         String url = "jdbc:hive2://server201:10000/one";
10         Connection con = DriverManager.getConnection(url,"hadoop","");
11         Statement st = con.createStatement();
12         ResultSet rs = st.executeQuery("select * from stud");
13         while(rs.next()) {
14             String id = rs.getString("id");
15             String name = rs.getString("name");
16             System.out.println("id:"+id+",name:"+name);
17         }
18         rs.close();
19         st.close();
20         con.close();
21     }
22 }
```

直接在IDEA中执行上述代码，将会在控制台输出查询的结果。

同样地，也可以在Hive的JDBC中执行聚合函数，同样会在Hadoop上执行一个MapReduce函数，虽然在IDEA中没有输出MapReduce的执行过程，但执行完成以后，通过查询8088端口，即可以看到刚才执行的MapReduce函数。以下是Hive JDBC执行count的代码，仅是将SQL语句修改为count(1)。

代码 6.4　HiveJdbcCount.java

```java
01 package org.hadoop.hive;
02 import java.sql.Connection;
03 import java.sql.DriverManager;
04 import java.sql.ResultSet;
05 import java.sql.Statement;
06 public class HiveJdbcCount {
07     public static void main(String[] args) throws Exception {
08         Class.forName("org.apache.hive.jdbc.HiveDriver");
09         String url = "jdbc:hive2://server201:10000/one";
10         Connection con = DriverManager.getConnection(url,"hadoop","");
11         Statement st  = con.createStatement();
12         ResultSet rs = st.executeQuery("select count(1) from stud");
13         rs.next();
14         int anInt = rs.getInt(1);
15         System.out.println("统计结果: "+anInt);
16         rs.close();
17         st.close();
18         con.close();
19     }
20 }
```

6.12　小　　结

- Hive是保存在HDFS上的数据库。Hive是HDFS的一个客户端。
- schematool命令用于初始化Hive的数据库，通过dbType指定数据库类型，如Derby或MySQL。
- 可以将HDFS数据映射成一张表，对这个表的操作就是对HDFS数据文件的分析。
- Hive的数据对应的是HDFS的文件，Hive的数据库对应的是HDFS的目录。
- Hive执行的语句被称为HQL，即Hive Query Language。
- load data local inpath是把Linux文件系统中的文件导入到Hive表中。
- load data inpath是把HDFS上的文件移动到Hive表所在的目录下。
- Hive默认使用Derby数据库作为matedata。可以通过配置将matedata修改成MySQL，这样就能支持多用户同时使用同一个matedata了。
- Hive拥有丰富的内置函数。可以使用show functions查看所有函数，使用"desc function 函数名;"查看具体某个函数的使用。
- 可以自定义Hive函数，即UDF开发。
- hiveserver2启动后，默认将占用10000端口，并可以通过JDBC连接。

第 7 章

HBase数据库实战

主要内容：

- ❖ HBase的特点。
- ❖ HBase的存储结构。
- ❖ HBase操作命令。
- ❖ HBase伪分布式。
- ❖ HBase分布式。

HBase是Hadoop DataBase的意思。HBase是一种构建在HDFS之上的分布式、面向列的存储系统。在需要实时读写、随机访问超大规模数据集时，可以使用HBase。

HBase是Google BigTable的开源实现，与Google BigTable利用GFS作为其文件存储系统类似，HBase利用Hadoop HDFS作为其文件存储系统，也是利用HDFS实现分布式存储的。Google运行MapReduce来处理BigTable中的海量数据，HBase同样利用Hadoop MapReduce来处理HBase中的海量数据；Google BigTable利用Chubby作为协同服务，HBase则利用ZooKeeper作为协同服务。

7.1 HBase 的特点

1. 大

一个表可以有上亿行，上百万列。

2. 面向列

面向列表（族）的存储和权限控制，列（族）独立检索。

3. 稀疏

对于为空（NULL）的列，并不占用存储空间，因此表可以设计得非常稀疏。

4. 无模式

每一行都有一个可以排序的主键和任意多的列，列可以根据需要动态增加，同一张表中不同的行可以有截然不同的列。

5. 数据多版本

每个单元中的数据可以有多个版本，默认情况下版本号自动分配，版本号就是单元格插入时的时间戳。

6. 数据类型单一

HBase中的数据都是字符串，没有类型。

（1）HBase的高并发和实时处理数据

Hadoop是一个高容错、高延时的分布式文件系统和高并发的批处理系统，虽然Hadoop MapReduce不适合用于提供实时计算；但HBase作为一种分布式数据库却能够提供实时计算，数据被保存在HDFS分布式文件系统上，由HDFS保证其高容错性，但是在生产环境中，HBase是如何基于Hadoop提供实时性的呢？HBase上的数据是以StoreFile（HFile）二进制流的形式存储在HDFS上Block块中的；但是HDFS并不知道HBase存储的是什么，它只把存储文件视为二进制文件，也就是说，HBase的存储数据对于HDFS文件系统是透明的。

图7-1是HBase文件在HDFS中的存储示意图。

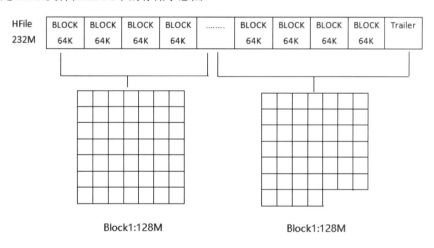

图7-1　HBase文件在HDFS中的存储示意图

HBase HRegionServer集群中所有的region数据在服务器启动时都是被打开的，并且在内存初始化一些memstore，相应地，这就在一定程度上加快了系统响应。而Hadoop中Block的数据文件默认是关闭的，只有在需要的时候才打开，处理完数据后就关闭，这在一定程度上增加了响应时间。

（2）HBase的数据模型

HBase以表的形式存储数据，表由行和列组成，列划分为若干个列族（Column Family），如图7-2所示。

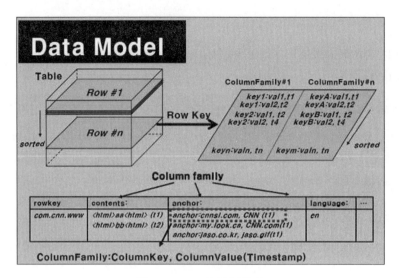

图7-2　HBase的数据模型

HBase的逻辑数据模型如表7-1所示。

表7-1　HBase 的逻辑数据模型

Row Key	Time Stamp	Column:"info"	Column:"other"		column:"..."
"key001"	t9		"other:name"	"Jerry"	
	t8		"other:age"	"100"	
	t6	"<HTML>..."			"col:value"
	t5	"Text ..."			
	t3	"Other..."			

① RowKey（行键）

与NoSQL数据库一样，RowKey是用来检索记录的主键。访问HBase table中的行只有三种方式：

- 通过单个RowKey访问。
- 通过RowKey的range全表扫描。
- RowKey可以使任意字符串（最大长度为64KB，实际应用中长度一般为10~100Bytes），在HBase内部，RowKey保存为字节数组。

在存储时，数据按照RowKey的字典序（byte order）排序存储。设计Key时，要充分考虑存储这个特性，将经常一起读取的行存储到一起（位置相关性）。注意，字典序对int排序的结果是1,10,100,11,12,13,14,15,16,17,18,19,20,21,…,9,91,92,93,94,95,96,97,98,99。要保存整型的自然序，RowKey必须用0进行左填充。

行的一次读写是原子操作（不论一次读写多少列）。这个设计决策能够使用户很容易理解程序在对同一个行进行并发更新操作时的行为。

② 列族（Column Family）

HBase表中的每个列都归属于某个列族。列族是表的Schema的一部分（而列不是），必须在使用表之前定义。列名都以列族作为前缀，例如courses:history、courses:math都属于courses这个列族。

③ 时间戳

HBase中通过Row和Columns确定的一个存储单元称为Cell。每个Cell都保存着同一份数据的多个版本。版本通过时间戳来索引，时间戳的类型是64位整型。时间戳可以由HBase（在数据写入时自动）赋值，此时时间戳是精确到毫秒的当前系统时间。时间戳也可以由客户显式赋值。如果应用程序要避免数据版本冲突，就必须自己生成具有唯一性的时间戳。每个Cell中，不同版本的数据按照时间戳倒序排序，即最新的数据排在最前面。

为了避免数据存在过多版本造成的管理（包括存储和索引）负担，HBase提供了两种数据版本回收方式。一是保存数据的最后n个版本，二是保存最近一段时间内的版本（比如最近七天）。用户可以针对每个列族进行设置。

④ Cell

Cell是由{row key, column(=< family> + < label>), version} 唯一确定的单元。Cell中的数据是没有类型的，全部是以字节码形式存储。

7.2　HBase 安装

HBase可以独立运行在一台主机上。此时，HBase将只会有一个进程，即HMaster（这个HMaster内含HRegionServer和HQuorumPeer两个服务）。虽然只有一个进程，但也可以提供HBase的大部分功能。

首先需要了解HBase的版本信息，在笔者完成本书时，HBase的最高版本为2.4，我们可以在HBase的官方网站上通过HBase documents查看HBase对JDK、Hadoop版本的兼容性。首先查看HBase对JDK版本的支持，如表7-2所示。

表7-2　HBase 对 JDK 版本的支持

Java Version	HBase 1.4+	HBase 2.2+	HBase 2.3+
JDK 7	√	×	×
JDK 8	√	√	√
JDK 11	×	×	*

说明：× 表示不支持，√表示支持，*表示未测试。

通过上表可以看出，任何版本都支持JDK 1.8+。而我们之前安装的环境，也正是基于JDK 1.8版本的。

我们再来看一下HBase对Hadoop版本的支持，如图7-3所示。

基于上面的配置，我们本次示例将选择HBase 2.3。现在就动手下载HBase，下载的地址为：https://mirrors.tuna.tsinghua.edu.cn/apache/hbase/2.3.4/hbase-2.3.4-bin.tar.gz。

	HBase-1.4.x	HBase-1.6.x	HBase-1.7.x	HBase-2.2.x	HBase-2.3.x
Hadoop-2.7.0	✗	✗	✗	✗	✗
Hadoop-2.7.1+	✓	✗	✗	✗	✗
Hadoop-2.8.[0-2]	✗	✗	✗	✗	✗
Hadoop-2.8.[3-4]	ⓘ	✗	✗	✗	✗
Hadoop-2.8.5+	ⓘ	✓	✗	✗	✗
Hadoop-2.9.[0-1]	✗	✗	✗	✗	✗
Hadoop-2.9.2+	ⓘ	✓	✓	✗	✗
Hadoop-2.10.x	ⓘ	✓	✓	ⓘ	✓
Hadoop-3.1.0	✗	✗	✗	✗	✗
Hadoop-3.1.1+	✗	✗	✗	✓	✓
Hadoop-3.2.x	✗	✗	✗	✓	✓

图7-3　HBase对Hadoop版本的支持

7.2.1　HBase 的单节点安装

HBase单节点运行方式，可以方便我们快速学习HBase的基本使用。HBase的单节点安装不需要Hadoop，数据保存到指定的磁盘目录下即可。在hbase-site.xml文件中，通过hbase.rootdir=file:///来指定保存的目录，也不需要ZooKeeper。启动HBase后，HBase将使用自带的ZooKeeper，只有一个进程HMaster（内部包含：HRegionServer和HQuorumPeerman两个子线程）。以下是HBase单节点安装的过程。

步骤01 上传并解压HBase。

同样地，使用tar将HBase解压到/app/目录下：

```
$ tar -zxvf hbase-2.3.4-bin.tar.gz -C /app/
```

步骤02 修改配置文件。

首先修改的是hbase-env.sh文件，此文件中保存了JAVA_HOME环境变量信息，配置如下：

```
$ vim /app/hbase-2.3.4/conf/hbase-env.sh
export JAVA_HOME=/usr/java/jdk1.8.0_281
export HBASE_MANAGES_ZK=true
```

再修改hbase-site.xml配置文件，此配置文件中，我们需要指定HBase数据的保存目录，由于目前是快速入门，所以可以先将HBase的数据保存到磁盘上。具体的配置如下：

```
$ vim /app/hbase-2.3.4/conf/hbase-site.xml
<configuration>
    <property>
        <name>hbase.rootdir</name>
        <value>file:///app/datas/hbase</value>
    </property>
    <property>
```

```xml
        <name>hbase.zookeeper.property.dataDir</name>
        <value>/app/datas/zookeeper</value>
    </property>
    <!--以下配置是否检查流功能 -->
    <property>
        <name>hbase.unsafe.stream.capability.enforce</name>
        <value>false</value>
    </property>
</configuration>
```

修改regionservers文件，此文件用于指定HRegionServer的服务器地址，由于是单机部署，所以指定本机名称即可。添加本机主机名：

```
$ vim regionservers
server201
```

步骤03 启动/停止HBase。

启动HBase只需要在HBase的bin目录下执行start-hbase.sh即可：

```
[hadoop@server201 app]$ hbase-2.3.4/bin/start-hbase.sh
SLF4J: Class path contains multiple SLF4J bindings.
SLF4J: Actual binding is of type [org.slf4j.impl.Log4jLoggerFactory]
```

启动完成以后，通过jps命令查看进程，会发现HMaster进程已经运行了：

```
[hadoop@server201 app]$ jps
7940 Jps
7785 HMaster
```

停止HBase只需要执行stop-hbase.sh即可：

```
[hadoop@server201 app]$ hbase-2.3.4/bin/stop-hbase.sh
stopping hbase....
```

步骤04 登录HBase Shell。

在bin目录下，使用hbase shell命令登录HBase的Shell，即可操作HBase数据库。登录成功后，将显示HBase命令行。

```
[hadoop@server201 app]$ hbase-2.3.4/bin/hbase shell
HBase Shell
Use "help" to get list of supported commands.
Use "exit" to quit this interactive shell.
For Reference, please visit: http://hbase.apache.org/2.0/book.html#shell
Version 2.3.4, rafd5e4fc3cd259257229df3422f2857ed35da4cc, Thu Jan 14 21:32:25 UTC 2021
Took 0.0006 seconds
hbase(main):001:0>
```

步骤05 HBase数据操作。

现在让我们快速创建一个表，保存一些数据，以了解HBase是如何保存数据的。首先，不了解HBase命令的读者，可以直接在HBase命令行输入help，此命令将会显示HBase的帮助信息。

创建一个命名空间，可以理解成创建了一个数据库：

```
hbase(main):009:0> create_namespace 'ns1'
Took 0.1776 seconds
```

查看所有命名空间，可以理解成查看所有数据库：

```
hbase(main):010:0> list_namespace
NAMESPACE
Default
hbase
ns1
3 row(s)
Took 0.0158 seconds
```

创建一个表，在指定的命名空间下，并指定列族为f。可以理解为创建一个表，并指定一个列名为f：

```
hbase(main):011:0> create 'ns1:stud','f'
Created table ns1:stud
Took 0.8601 seconds
=> Hbase::Table - ns1:stud
```

向表中写入一行记录，其中R001为主键，即Row Key，f:name为列名，Jack为列值。此处与关系数据库有很大的区别，注意区分：

```
hbase(main):012:0> put 'ns1:stud','R001','f:name','Jack'
Took 0.3139 seconds
hbase(main):013:0> put 'ns1:stud','R002','f:age','34'
Took 0.0376 seconds
```

查询我们创建的表，语句类似于关系数据库中的select：

```
hbase(main):014:0> scan 'ns1:stud'
ROW                     COLUMN+CELL
 R001                   column=f:name, timestamp=1568092417316, value=Jack
 R002                   column=f:age, timestamp=1568092435076, value=34
2 row(s)
Took 0.0729 seconds
```

7.2.2 HBase 的伪分布式安装

HBase的伪分布式安装，即将HBase的数据保存到伪分布式的HDFS系统中。此时Hadoop环境为伪分布式，HBase的节点只有一个，HBase使独立运行的ZooKeeper。以下步骤将配置一个HBase的伪分布式运行环境，它的要求如下：

（1）配置好的Hadoop运行环境。
（2）独立运行的ZooKeeper，只有一个节点。
（3）将Hadoop的ZooKeeper指向这个独立的ZooKeeper。

步骤01 准备Hadoop和ZooKeeper。

请根据前面章节的步骤，配置好Hadoop伪分布式运行环境，同时也配置好ZooKeeper的独立节点运行环境。

测试以上两个环境正常运行并可用。

步骤02 修改HBase配置文件。

首先修改HBase的配置文件——hbase-env.sh文件，在此配置文件中，重点配置ZooKeeper选项，配置为使用外部的ZooKeeper即可。具体配置如下：

```
$ vim /app/hbase-2.3.4/conf/hbase-env.sh
export JAVA_HOME=/usr/java/jdk1.8.0-281
export HBASE_MANAGES_ZK=false
```

修改HBase配置文件——hbase-site.xml文件，重点关注hbase.rootdir用于指定在Hadoop中存储HBase数据的目录。hbase.zookeeper.quorum用于指定ZooKeeper地址。

```xml
<configuration>
    <property>
        <name>hbase.cluster.distributed</name>
        <value>true</value>
    </property>
    <property>
        <name>hbase.tmp.dir</name>
        <value>/app/datas/hbase/tmp</value>
    </property>
    <property>
        <name>hbase.unsafe.stream.capability.enforce</name>
        <value>false</value>
    </property>
    <property>
        <name>hbase.rootdir</name>
        <value>hdfs://server201:8020/hbase</value>
    </property>
    <property>
        <name>hbase.zookeeper.quorum</name>
        <value>server201:2181</value>
    </property>
</configuration>
```

步骤03 启动HBase。

直接使用start-hbase.sh即可启动HBase。

```
$ /app/hbase-2.3.4/bin/start-hbase.sh
```

查看HBase的进程，请注意以下黑体标记的进程：

```
[hadoop@server201 conf]$ jps
2992 NameNode
3328 SecondaryNameNode
5793 Jps
5186 Main
3560 ResourceManager
4120 QuorumPeerMain
4728 HRegionServer
3691 NodeManager
4523 HMaster
3119 DataNode
```

同样地，可以通过浏览器查看16010端口，如图7-4所示。

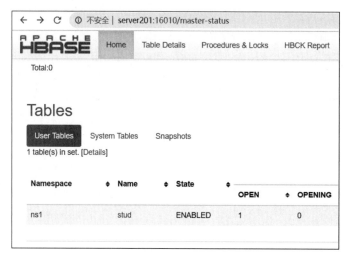

图7-4 查看16010端口

7.2.3 Java 客户端代码

HBase分布式启动成功后，就可以通过Java代码来操作HBase中的数据。

使用Java代码操作需要添加两个依赖：一个是Hadoop的，因为HBase依赖的某些配置对象是Hadoop包中的；另一个是hbase-client依赖。具体依赖如下：

```xml
<dependency>
    <groupId>org.apache.hbase</groupId>
    <artifactId>hbase-client</artifactId>
    <version>2.3.4</version>
</dependency>
<dependency>
    <groupId>org.apache.hadoop</groupId>
    <artifactId>hadoop-client</artifactId>
    <version>3.2.2</version>
</dependency>
```

代码7.1用于测试与HBase的连接。注意，我们连接的端口为ZooKeeper的端口2181。

代码 7.1 ConnectTest.java

```
01 package org.hadoop.hbase;
02 import lombok.extern.slf4j.Slf4j;
03 import org.apache.hadoop.conf.Configuration;
04 import org.apache.hadoop.hbase.HBaseConfiguration;
05 import org.apache.hadoop.hbase.NamespaceDescriptor;
06 import org.apache.hadoop.hbase.client.Admin;
07 import org.apache.hadoop.hbase.client.Connection;
08 import org.apache.hadoop.hbase.client.ConnectionFactory;
09 @Slf4j
10 public class ConnectTest {
11     public static void main(String[] args) throws Exception {
12         Configuration configuration = HBaseConfiguration.create();
13         configuration.set("hbase.zookeeper.property.clientPort", "2181");
14         configuration.set("hbase.zookeeper.quorum", "server201");
15         Connection connection = ConnectionFactory.createConnection(configuration);
```

```
16         log.info("测试连接: "+connection);
17         Admin admin = connection.getAdmin();
18         NamespaceDescriptor[] ns = admin.listNamespaceDescriptors();
19         log.info("命名空间个数: "+ns.length);
20         for(NamespaceDescriptor n:ns){
21             System.out.println("命名空间: "+n.getName());
22         }
23         connection.close();
24     }
25 }
```

测试输出结果：

测试连接：hconnection-0x131774fe
命名空间个数：3
命名空间：default
命名空间：hbase
命名空间：ns1

为了给读者快速展示更多的示例，我们通过一个Junit测试来展示更多HBase的操作，这些操作中包含查询表名称、查询表中的数据等。

代码 7.2 HBaseOperation.java

```
01 package org.hadoop.hbase;
02 import lombok.extern.slf4j.Slf4j;
03 import org.apache.hadoop.conf.Configuration;
04 import org.apache.hadoop.hbase.Cell;
05 import org.apache.hadoop.hbase.CellUtil;
06 import org.apache.hadoop.hbase.HBaseConfiguration;
07 import org.apache.hadoop.hbase.TableName;
08 import org.apache.hadoop.hbase.client.*;
09 import org.apache.hadoop.hbase.util.Bytes;
10 import org.junit.After;
11 import org.junit.Before;
12 import org.junit.Test;
13 import java.util.List;
14 @Slf4j
15 public class HBaseOperation {
16     private Connection con;
17     @Before
18     public void before() throws Exception {
19         Configuration configuration = HBaseConfiguration.create();
20         configuration.set("hbase.zookeeper.property.clientPort", "2181");
21         configuration.set("hbase.zookeeper.quorum", "server201");
22         con = ConnectionFactory.createConnection(configuration);
23     }
24     @After
25     public void after() throws Exception {
26         con.close();
27     }
28     /**
29      * 查询所有表
30      */
31     @Test
32     public void listTables() throws Exception {
33         Admin admin = con.getAdmin();
```

```java
34          List<TableDescriptor> list = admin.listTableDescriptors();
35          for (TableDescriptor t : list) {
36              log.info("tableName:" + t.getTableName().getNameAsString());
37          }
38     }
39     /**
40      * 查询指定命名空间的表
41      */
42     @Test
43     public void listNamespaceTables() throws Exception {
44         Admin admin = con.getAdmin();
45         TableName[] tns = admin.listTableNamesByNamespace("ns1");
46         for (TableName tn : tns) {
47             log.info("name :" + tn.getNameAsString());
48         }
49         con.close();
50     }
51     /**
52      * 查询表中的数据
53      */
54     @Test
55     public void queryData() throws Exception {
56         Table table = con.getTable(TableName.valueOf("ns1:stud"));
57         Scan scan1 = new Scan();
58         ResultScanner rs = table.getScanner(scan1);
59         for (Result r : rs) {
60             for (Cell cell : r.listCells()) {
61                 log.info("RowKey:" + Bytes.toString(CellUtil.cloneRow(cell)) + "\t"
62                         + Bytes.toString(CellUtil.cloneFamily(cell)) + "\t"
63                         + Bytes.toString(CellUtil.cloneQualifier(cell)) + "\t"
64                         + Bytes.toString(CellUtil.cloneValue(cell)));
65             }
66             log.info("--------------");
67         }
68         rs.close();
69     }
70     /**
71      * 创建一个表
72      */
73     @Test
74     public void createTable() throws Exception {
75         Admin admin = con.getAdmin();
76         TableDescriptor td =
77             TableDescriptorBuilder
78                     .newBuilder(TableName.valueOf("ns1:person"))//表名
79                     .setColumnFamily(ColumnFamilyDescriptorBuilder.of("c"))  //列族
80                     .build();//创建TableDescriptor
81         admin.createTable(td);//创建Table
82     }
83     /**
84      * 删除一个表
85      */
86     @Test
87     public void dropTable() throws Exception {
88         Admin admin = con.getAdmin();
89         admin.disableTable(TableName.valueOf("ns1:person"));
90         admin.deleteTable(TableName.valueOf("ns1:person"));
91         admin.close();
```

```
 92      }
 93      /**
 94       * 向表中插入数据：（如果key值已经存在，则为修改）
 95       */
 96      @Test
 97      public void putData() throws Exception {
 98          Table table = con.getTable(TableName.valueOf("ns1:stud"));
 99          Put put = new Put("K002".getBytes());
100          put.addColumn("f".getBytes(), "name".getBytes(), "Rose".getBytes());
101          table.put(put);
102          table.close();
103      }
104      /**
105       * 根据RowKey删除记录
106       */
107      @Test
108      public void deleteByRowKey() throws Exception {
109          Table table = con.getTable(TableName.valueOf("ns1:stud"));
110          Delete delete = new Delete(Bytes.toBytes("K002"));
111          table.delete(delete);
112          table.close();
113      }
114      /**
115       * 删除一个单元格，则需要在Delete中输入一个Column即可
116       */
117      @Test
118      public void deleteColumn() throws Exception {
119          String key = "K001";
120          Table table = con.getTable(TableName.valueOf("ns1:stud"));
121          boolean boo = table.exists(new Get(Bytes.toBytes(key)));
122          if (boo) {
123              Delete delete = new Delete(Bytes.toBytes(key));
124              //设置需要删除的列，即可以只删除一个单元格
125              delete.addColumn(Bytes.toBytes("f"), Bytes.toBytes("name"));
126              table.delete(delete);
127          } else {
128              log.info("此rowkey不存在:" + key);
129          }
130          table.close();
131      }
132 }
```

还有其他更多的操作，请参考下面的讲解。

(1) 通过RowKey查询

只要在Scan中传递Get对象，即可通过RowKey查询：

```
Scan scan = new Scan(new Get("U001".getBytes()));
```

(2) 根据值来进行查询

根据值来查询，使用ValueFilter对象，BinaryComparator用于比较二进制数据：

```
Scan scan = new Scan();
//设置查询列的信息
BinaryComparator bc = new BinaryComparator("Mary".getBytes());
ValueFilter vf = new ValueFilter(CompareOp.EQUAL, bc);
```

```
scan.setFilter(vf);
```

(3)使用正则表达式的查询

RegexStringComparator用于执行正则表达式的查询：

```
RegexStringComparator bc = new RegexStringComparator(".*r.*");
ValueFilter vf = new ValueFilter(CompareOp.EQUAL, bc);
scan.setFilter(vf);
```

或使用SingleColumnValueFilter指定查询的列名：

```
RegexStringComparator bc = new RegexStringComparator(".*r.*");
SingleColumnValueFilter vf =
        new SingleColumnValueFilter(
            Bytes.toBytes("info"),
            Bytes.toBytes("name"),
            CompareOp.EQUAL , bc);
scan.setFilter(vf);
```

(4)字符串包含查询

可以使用SubStringComparator查询包含的字符串：

```
ByteArrayComparable bc =new SubstringComparator("Mary");
```

(5)前缀二进制比较器

BinaryPrefixComparator是前缀二进制比较器。与二进制比较器不同的是，它只比较前缀是否相同。以下查询info:name列以Ma为前缀的数据。

```
Scan scan = new Scan();
BinaryPrefixComparator comp
    = new BinaryPrefixComparator(Bytes.toBytes("Ma"));
SingleColumnValueFilter filter
    = new SingleColumnValueFilter(Bytes.toBytes("info"),
        Bytes.toBytes("name"), CompareOp.EQUAL, comp);
scan.setFilter(filter);
```

(6)列值过滤器

SingleColumnValueFilter用于测试值的情况为相等、不等、范围，等等。

下面示例检测列族family下列qualifier的列值和字符串"some-value"相等的部分：

```
Scan scan = new Scan();
SingleColumnValueFilter filter
        = new SingleColumnValueFilter(Bytes.toBytes("family"),
            Bytes.toBytes("qualifier"),
            CompareOp.EQUAL, Bytes.toBytes("some-value"));
scan.setFilter(filter);
```

(7)排除过滤

SingleColumnValueExcludeFilter跟 SingleColumnValueFilter 功能一样，只是不查询出该列的值。下面代码就不会查询出family列族下qualifier列的值：

```
Scan scan = new Scan();
SingleColumnValueExcludeFilter filter
    = new SingleColumnValueExcludeFilter(Bytes.toBytes("family"),
```

```
            Bytes.toBytes("qualifier"),
CompareOp.EQUAL, Bytes.toBytes("some-value"));
scan.setFilter(filter);
```

(8)列族过滤器

FamilyFilter用于过滤列族(通常在Scan过程中通过设定某些列族来实现该功能,而不是直接使用该过滤器):

```
Scan scan = new Scan();
FamilyFilter filter
= new FamilyFilter(CompareOp.EQUAL,
        new BinaryComparator(Bytes.toBytes("some-family")));
scan.setFilter(filter);
```

(9)列名过滤器

QualifierFilter用于列名(Qualifier)过滤:

```
QualifierFilter qff =
            new QualifierFilter(CompareOp.EQUAL,
            new BinaryComparator("name".getBytes()));
            scan.setFilter(qff);
```

(10)列名前缀过滤器

ColumnPrefixFilter用于列名(Qualifier)前缀过滤,即包含某个前缀的所有列名:

```
Scan scan = new Scan();
ColumnPrefixFilter filter =
    new ColumnPrefixFilter(Bytes.toBytes("somePrefix"));
scan.setFilter(filter);
```

(11)多个列名前缀过滤器

MultipleColumnPrefixFilter 与 ColumnPrefixFilter的行为类似,但它还可以指定多个列名(Qualifier)前缀:

```
Scan scan = new Scan();
byte[][] prefixes = new byte[][]{
Bytes.toBytes("prefix1"),
Bytes.toBytes("prefix2")};
MultipleColumnPrefixFilter filter =
new MultipleColumnPrefixFilter(prefixes); scan.setFilter(filter);
```

(12)列范围过滤器

ColumnRangeFilter过滤器可以进行高效的列名内部扫描:

```
Scan scan = new Scan();
boolean minColumnInclusive = true;
boolean maxColumnInclusive = true;
ColumnRangeFilter filter =
new ColumnRangeFilter(
        Bytes.toBytes("minColumnName"), minColumnInclusive,
        Bytes.toBytes("maxColumnName"), maxColumnInclusive);
scan.setFilter(filter);
```

（13）行键过滤器

RowFilter行键过滤器，一般来讲，执行Scan使用startRow/stopRow方式比较好，而RowFilter过滤器也可以完成对某一行的过滤：

```
Scan scan = new Scan();
RowFilter filter =
        new RowFilter(CompareOp.EQUAL,
        new BinaryComparator(Bytes.toBytes("someRowKey1")));
scan.setFilter(filter);
```

（14）分页过滤器

PageFilter用于按行分页。必须设置每次显示几行，以及在Scan中设置开始的行值：

```
Scan scan = new Scan();
PageFilter pf = new PageFilter(5);
scan.setFilter(pf);
byte[] startRow = Bytes.add("U005".getBytes(),Bytes.toBytes("postfix"));
scan.setStartRow(startRow);
ResultScanner resultScanner = table.getScanner(scan);
```

（15）串联多个过滤器

可以使用FilterList将多个过滤器串联起来，组成And或Or的过滤器：

```
FilterList fl = new FilterList(Operator.MUST_PASS_ALL);
fl.addFilter(new ValueFilter(CompareOp.EQUAL,
        new BinaryComparator("Smith".getBytes())));
fl.addFilter(new RowFilter(CompareOp.EQUAL,
        new BinaryComparator("U005".getBytes())));
scan.setFilter(fl);
```

由于过滤器比较多，更多的过滤器用法将不再一一展示，请读者自行查看HBase官方文档。

7.3 HBase 集群安装

在HBase集群环境下，会有一个HMaster进程和多个HRegionServer进程，其中HRegionServer会根据所配置的节点情况运行在多台主机上。本节将带领读者安装一个HBase的集群环境。在安装时，要注意HBase与Hadoop的兼容关系。我们采用的版本如下：

Hadoop-3.2.2

HBase-2.3.4

ZooKeeper-3.6.2

三台服务器的配置如表7-3所示。

表 7-3 三台服务器的配置

Ip/name	软件	进程
192.168.56.101 server101	Hadoop-3.2.2 ZooKeeper-3.6.2 HBase-2.3.4	NameNode SecondaryNameNode ResourceManager NodeManager DataNode QuorumPeerMan HMaster HRegionServer
192.168.56.102 server102	Hadoop-3.2.2 ZooKeeper-3.6.2 HBase-2.3.4	DataNode NodeManager QuorumPeerMan HRegionServer
192.168.56.103 server103	Hadoop-3.2.2 ZooKeeper-3.6.2 HBase-2.3.4	DataNode NodeManager QuorumPeerMan HRegionServer

安装ZooKeeper集群和Hadoop集群的过程请参考之前的章节，下面直接安装HBase。

步骤01 上传并解压HBase。

```
$tar -zxvf ~/hbase-2.3.4-bin.tar.gz  -C  /app/
```

目录名称太长，修改一下名称：

```
$ mv /app/hbase-2.3.4-bin /app/hbase-2.3.4
```

步骤02 配置环境变量。

添加HBase环境变量：

```
$ sudo vim /etc/profile
export HBASE_HOME=/app/hbase-2.3.4
export PATH=$PATH:$HBASE_HOME/bin
```

让环境变量生效：

```
$ source /etc/profile
```

步骤03 配置HBase。

修改hbase-env.sh文件：

```
export JAVA_HOME=/usr/java/jdk1.8.0_281
export HBASE_MANAGES_ZK=false
```

配置hbase-site.sh文件：

```
<configuration>
    <property>
        <name>hbase.tmp.dir</name>
        <value>/app/datas/hbase/tmp</value>
```

```xml
        </property>
        <property>
            <name>hbase.rootdir</name>
            <value>hdfs://server201:8020/hbase</value>
        </property>
        <property>
            <name>hbase.cluster.distributed</name>
            <value>true</value>
        </property>
        <property>
            <name>hbase.zookeeper.quorum</name>
            <value>server101:2181,server102:2181,server103:2181</value>
        </property>
        <property>
            <name>hbase.unsafe.stream.capability.enforce</name>
            <value>false</value>
        </property>
</configuration>
```

- hbase.rootdir：用于配置HBase的文件系统所保存的目录。
- hbase.tmp.dir：保存HBase临时文件的目录，如果不设置，将保存到/tmp目录下。
- hbase.zookeeper.property.dataDir：用于配置HBase自带的ZooKeeper数据文件保存的目录。
- hbase.zookeeper.quorum：用于设置外部ZooKeeper的地址和目录。需要在hbase-env.sh中将ZooKeeper设置为false。

配置RegionServer：

```
server101
server102
server103
```

将所有关于HBase的配置分发到其他主机，包含环境变量配置文件和HBase整个目录。

```
$ scp /etc/profile root@server102:/etc/
$ scp /etc/profile root@server103:/etc/
$ scp -r /app/hbase-2.3.4 server102:/app/
$ scp -r /app/hbase-2.3.4 server103:/app/
```

步骤04 启动HBase。

与之前一样，执行$HBASE_HOME/bin目录下的start-hbase.sh即可以启动HBase。由于是集群配置，请先启动Hadoop和ZooKeeper。执行start-hbase.sh后，将会根据配置启动其他节点上的HRegionServer。

```
$ /app/hbase-2.3.4/bin/start-hbase.sh
```

jps查看进程：

```
[hadoop@server101 app]$ jps
9073 ResourceManager
22433 HRegionServer
5442 DataNode
23075 Jps
9205 NodeManager
5676 SecondaryNameNode
17357 QuorumPeerMain
5310 NameNode
```

```
22270 HMaster
```

查看其他节点的进程，如server102主机上的进程：

```
[hadoop@server102 app]$ jps
22433 HRegionServer
5446 DataNode
23015 Jps
9203 NodeManager
17337 QuorumPeerMain
```

步骤 05 访问16010端口。

使用浏览器访问16010端口，将会看到HBase整个集群的情况，如图7-5所示。

如果要启动高可用，则可以在server102，即另一台机器上再启动一个master，命令如下：

```
$ /app/hbase-2.3.4/bin/hbase-deamon.sh start master
```

启动以后，访问server102上的16010端口，此时server102会显示为Backup Master，即为备份节点，如图7-6所示。

图7-5 查看HBase集群

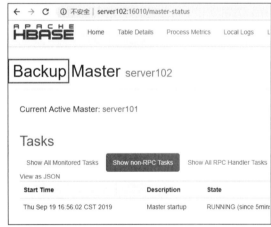

图7-6 显示备份节点

注意：
- 启动HBase之前，必须确定已经在/etc/hosts文件中配置了本地DNS。
- 使用http://ip:16010访问之前，必须确定防火墙已经关闭。关闭防火墙使用sudo systemctl stop firewalld.service命令。禁用防火墙使用sudo systemctl disable firewalld.service命令。

7.4 HBase Shell 操作

HBase Shell提供了大多数的HBase命令，通过HBase Shell，用户可以方便地创建、删除及修改表，还可以向表中添加数据、罗列出表中的相关信息等。在启动HBase之后，用户可以通过下面命令进入HBase Shell命令行模式，HBase脚本文件在$HBASE_HOME/bin/目录下。

```
$./hbase shell
```

在登录成功HBase Shell以后，可以使用help显示所有命令列表：

```
hbase(main):002:0> help
```

结果可以看到命令分组，如表7-4所示（注意大部分命令都是小写，少部分有大小写的区分）。

表7-4　HBase 命令

Group Name	Commands
general	status,version
ddl	alter,create,describe,disable,drop,enable,exists,is_disable,is_enable,list
dml	count,delete,deleteall,get_counter,incr,put,scan,truncate
tools	assign,blance_switch,blancer,close_region,compact,flush,major_compact,move,split,unassign,...
replication	add_peer, append_peer_tableCFs, disable_peer, disable_table_replication, enable_peer,...
...	...

由于命令比较多，请读者自行查看。下面将演示一些常用命令的操作。

7.4.1　数据模型定义

（1）查看版本version

```
hbase(main):008:0> version
2.3.4, r930b9a55528fe45d8edce7af42fef2d35e77677a, Thu Apr  6 19:36:54 PDT 2021
```

（2）查看状态status

```
hbase(main):009:0> status
1 active master, 0 backup masters, 1 servers, 1 dead, 2.0000 average load
```

还有一些DDL操作如create。

（3）创建一个表create

语法：create "表名","列族1","列族2","列族N"

或create "表名",{NAME=>"列族1",VERSIONS=>保存版本数量},{....}

命令中，保存版本数量指用于记录一个列族最多可以保存的历史记录。

```
hbase(main):005:0> create "stud","info"
0 row(s) in 1.4040 seconds
=> Hbase::Table - stud
hbase(main):006:0> list
TABLE
Stud
1 row(s) in 0.0290 seconds
=> ["stud"]
hbase(main):007:0> describe "stud"
```

上面的代码中，stud为表名，可以使用""""双引号，也可以使用"''"单引号。info为列族。还可以指定更多的信息，如使用VERSIONS来指定保存的版本信息：

```
hbase(main):008:0> create "person",{NAME=>"info",VERSIONS=>3}
0 row(s) in 1.2580 seconds
=> Hbase::Table - person
hbase(main):009:0> list
TABLE
person
stud
2 row(s) in 0.0250 seconds
=> ["person", "stud"]
```

person为表名，在person后面通过"{ }"大括号声明列族为info，版本信息为3。在创建表以后，通过list可以显示所有的数据表。

（4）修改表结构alter

添加一个新的列，可以使用alter语句：

语法：alter "表名",{NAME=>"列族名称",VERSIONS=>3}

如果修改的列族不存在，则会添加一个新的列族，否则会修改已有列族信息。在修改之前，可以先通过desc查看表的信息，在修改之后，再通过desc查看表的信息。

```
hbase(main):003:0> desc "stud"
```

查看表的信息，结果会列出这个表的所有列族信息：

```
hbase(main):003:0> desc "stud"
Table stud is ENABLED
stud
COLUMN FAMILIES DESCRIPTION
{NAME => 'info', BLOOMFILTER => 'ROW', VERSIONS => '1',...
```

通过上面的列族信息可以看出，info列族的VERSIONS为1，现在可以通过alter修改info列族的VERSIONS为3：

```
hbase(main):004:0> alter "stud",{NAME=>"info",VERSIONS=>3}
```

然后，再通过desc查看stud表的信息：

```
hbase(main):005:0> desc "stud"
Table stud is ENABLED
stud
COLUMN FAMILIES DESCRIPTION
{NAME => 'info', BLOOMFILTER => 'ROW', VERSIONS => '3',....
```

此时可以看到，VERSIONS已经修改为3了。也可以通过alter添加一个新的列族，如果这个列族不存在，则会创建一个新的列族：

```
hbase(main):006:0> alter "stud",{NAME=>"desc",VERSIONS=>3}
```

然后再通过desc查看stud表的结构：

```
hbase(main):007:0> desc "stud"
Table stud is ENABLED
```

```
stud
COLUMN FAMILIES DESCRIPTION
{NAME => 'desc', BLOOMFILTER => 'ROW', VERSIONS => '3', ...
{NAME => 'info', BLOOMFILTER => 'ROW', VERSIONS => '3', ....
```

从结果可见，已经添加了一个新的列族desc。

注意：HBase的列族信息是按字典顺序来排序的，所以上面例子中desc的列族信息在info列族信息的前面。

（5）删除一个列族

语法：alter "表名",{NAME=>"列族名称",METHOD=>"delete"}

删除一个列族，同样使用alter关键字，只是必须传递method=>"delete"来指定删除。

```
hbase(main):009:0> alter "stud",{NAME=>"desc",METHOD=>"delete"}
```

再次查看这个表的信息：

```
hbase(main):010:0> desc "stud"
{NAME => 'info', BLOOMFILTER => 'ROW', VERSIONS => '3',.....
```

可以看到，只剩下一个列族info了，desc列族已经被删除。

7.4.2 数据基本操作

DML操作用于向表中写入数据、查询数据及删除数据，功能类似于SQL语句的DML，但命令与SQL不同。

（1）插入数据

语法：put "表名","行键","列族:列名","具体值"

HBase写入的数据没有数据类型，都是二进制数据。相同的列族属于同一行数据。

向表中写入一行数据：

```
hbase(main):012:0> put "stud","U001","info:name","Jack"
```

上例中，stud为表名，U001为行键（即RowKey的值），info后面的name为列的名称，Jack为列值。在插入数据以后，就可以通过表描述显示表中的所有数据了。

（2）扫描表中的数据

语法：scan "表名"

scan用于显示表中的所有数据，类似于SQL语法中的select * from xxx。

```
hbase(main):013:0> scan "stud"
ROW    COLUMN+CELL
 U001    column=info:name, timestamp=1500782803718, value=Jack
1 row(s) in 0.0710 seconds
```

上例中最后一行说明目前只有一行数据。我们可以写入多行数据，然后再通过scan查看表数据。

再写入一行数据，行键与前面的行键相同：

```
hbase(main):014:0> put "stud","U001","info:age",23
```

然后再扫描表中的记录,依然是一行数据,因为一个行键代表的是一行数据。

```
hbase(main):015:0> scan "stud"
ROW    COLUMN+CELL
 U001    column=info:age, timestamp=1500783145216, value=23
 U001    column=info:name, timestamp=1500782803718, value=Jack
1 row(s) in 0.0280 seconds
```

再写入一个不同的行键:

```
hbase(main):016:0> put "stud","U002","info:name","Mary"
```

然后再查看表中的记录,已经发现有两行记录了,因为行键不同。

```
hbase(main):017:0> scan "stud"
ROW    COLUMN+CELL
 U001    column=info:age, timestamp=1500783145216, value=23
 U001    column=info:name, timestamp=1500782803718, value=Jack
 U002    column=info:name, timestamp=1500783317379, value=Mary
2 row(s) in 0.0280 seconds
```

可以通过put多次修改列的值,首先查看U001和info:name的值:

```
hbase(main):018:0> scan "stud"
ROW    COLUMN+CELL
 U001     column=info:name, timestamp=1500782803718, value=Jack
 ....
2 row(s) in 0.0520 seconds
```

上面info:name的值为Jack,以下分别修改N次:

```
hbase(main):019:0> put "stud","U001","info:name","FirstName"
hbase(main):020:0> put "stud","U001","info:name","SecondName"
hbase(main):021:0> put "stud","U001","info:name","ThirdName"
```

然后再扫描表,info:name值为最后一次修改的记录:

```
hbase(main):022:0> scan "stud"
ROW     COLUMN+CELL
 U001     column=info:name, timestamp=1500783878309, value=ThirdName
 ...
```

HBase的VERSIONS主要控制保存的版本记录,上面对列info:name的数据修改了3次,可以通过版本扫描,显示所有修改过的记录。

(3)扫描时显示各版本的记录

可以通过指定VERSIONS=>3显示最近三次修改记录,使用RAW显示操作信息。在修改数据时,每一次都会记录一个timestamp,即系统的当前时间。

```
hbase(main):034:0> scan "stud",{RAW=>true,VERSIONS=>3,COLUMNS=>"info"}
ROW     COLUMN+CELL
 U001   column=info:age, timestamp=1500783145216, value=23
 U001   column=info:name, timestamp=1500783878309, value=ThirdName
 U001   column=info:name, timestamp=1500783873566, value=SecondName
 U001   column=info:name, timestamp=1500783868549, value=FirstName
 .....
```

```
2 row(s) in 0.0480 seconds
```

通过上面的查询可以看到，VERSIONS=>3显示了最近三次操作的记录。timestamp为操作时间，以倒序显示。

（4）扫描过滤

也可以在扫描时使用过滤功能。如果指定从哪一个行键开始，则可以使用STARTROW过滤：

```
hbase(main):036:0> scan "stud",{COLUMNS=>"info",STARTROW=>"U002"}
ROW       COLUMN+CELL
 U002    column=info:name, timestamp=1500783317379, value=Mary
1 row(s) in 0.0240 seconds
```

也可以使用ENDROW指定结束的行键，但需要注意，ENDROW的值是不包含的，即如果要到行键的值为U003且包含U003，则应该指定ENDROW=>"U004"：

```
hbase(main):043:0>scan stud",{COLUMNS=>"info",STARTROW=>"U002",ENDROW=>"U004"}
ROW      COLUMN+CELL
 U002    column=info:name, timestamp=1500783317379, value=Mary
 U003    column=info:name, timestamp=1500784918046, value=Alex
2 row(s) in 0.0340 seconds
```

上面的显示的范围为>=U002且< U004。

（5）值过滤

可以使用ValueFilter实现值过滤功能。因为HBase中数据是以二进制开始保存的，所以比较方式为binary（即二进制）。如下示例查询值等于Mary的记录：

```
hbase(main):003:0> scan "stud",FILTER=>"ValueFilter(=,'binary:Mary')"
ROW        COLUMN+CELL
 U002    column=info:name, timestamp=1500783317379, value=Mary
```

结果显示值为Mary的记录，值得说明的是，如果多个列的值为Mary，则会查询出多列的数据。值比较与列名无关。

（6）包含

包含类似于contains，如果值中包含某个字符串，则可以查询出相关记录。以下示例是查询记录值里面包含Mary的记录：

```
hbase(main):005:0>scan \
"stud",FILTER=>"ValueFilter(=,'substring:Mary')"
    ROW    COLUMN+CELL
    U002  column=info:name, timestamp=1500783317379, value=Mary
    U004  column=info:desc, timestamp=1500790561908, value=Mary ..
```

（7）列名过滤

也可以使用ColumnPrefixFilter只查询某些指定的列。如下面示例显示所有name的列：

```
hbase(main):017:0> scan "stud",FILTER=>"ColumnPrefixFilter('name')"
ROW       COLUMN+CELL
 U001    column=info:name, timestamp=1500783878309, value=ThirdName
 U002    column=info:name, timestamp=1500783317379, value=Mary
 U003    column=info:name, timestamp=1500784918046, value=Alex
```

使用ColumnPrefixFilter时，过滤的是列族后面的列名，而不是列族的名称。

（8）多个条件进行组合

使用AND或OR关键字，可以串联多个过滤条件。如下面示例查询列名为name且值中包含Mary的记录。

```
hbase>scan stud",FILTER=>"ColumnPrefixFilter('name') AND ValueFilter(=,'substring:Mary')"
ROW     COLUMN+CELL
U002    column=info:name, timestamp=1500783317379, value=Mary
```

（9）行键过滤

可以使用PrefixFilter实现行键的过滤。如下面示例只查询行键以U001开始的记录：

```
hbase(main):020:0> scan "stud",FILTER=>"PrefixFilter('U001')"
ROW     COLUMN+CELL
U001    column=info:age, timestamp=1500783145216, value=23
U001    column=info:name, timestamp=1500783878309, value=ThirdName
```

（10）数据查询

语法：get "表名","行键"[,"列族:[列名]]

行键是必须存在的，列族和列名可以省略。

查询行键的值为U001的记录：

```
hbase(main):022:0> get "stud","U001"
COLUMN   CELL
 info:age   timestamp=1500783145216, value=23
 info:name  timestamp=1500783878309, value=ThirdName
```

只查询行键的值为U001且列族名称为info的记录：

```
hbase(main):023:0> get "stud","U001","info"
COLUMN    CELL
 info:age    timestamp=1500783145216, value=23
 info:name   timestamp=1500783878309, value=ThirdName
```

查询行键的值为U001且列的名称为info:name的记录：

```
hbase(main):024:0> get "stud","U001","info:name"
COLUMN  CELL
info:name  timestamp=1500783878309, value=ThirdName
```

（11）修改数据

通过put操作表中已经存在的数据，即为修改：

```
hbase(main):020:0> put "stud","rk001","info:name","Jerry"
```

（12）删除数据

语法：deleteall "表名","行键","列族:列名"

```
hbase(main):083:0> delete "stud","rk001","info:age"
```

（13）删除整个行键中的所有数据

语法：deleteall "表名","行键"

```
hbase(main):087:0> deleteall "stud","rk001"
0 row(s) in 0.0190 seconds
```

(14)删除整个表中的所有数据

语法：truncate "表名"

```
hbase(main):090:0> truncate "stud"
```

HBase还有更多的操作命令，在此就不再赘述了。读者有了上面的知识，完全可以通过查看HBase官方文档获取所有命令的使用方式。

7.5 协处理器

协处理器在HBase中可以帮助我们执行类似聚合的操作，如sum、avg等。协处理器有两种：Observer和Endpoint。

Observer类似于传统数据库中的触发器，当发生某些事件的时候，这类协处理器会被Server端调用。Observer Coprocessor就是一些散布在HBase Server端代码中的hook钩子，在固定的事件发生时被调用。比如：put操作之前有钩子函数prePut，该函数在put操作执行前会被RegionServer调用；在put操作之后则有postPut钩子函数。以下演示协处理器在Java代码中的使用。

使用协处理的方式有很多种，其中一种就是直接在hbase-site.xml文件添加相关配置，配置后将全局有效。在hbase-site.xml文件中添加以下配置：

```
<property>
    <name>hbase.coprocessor.user.region.classes</name>
    <value>org.apache.hadoop.hbase.coprocessor.AggregateImplementation</value>
</property>
```

在项目中添加以下依赖，在原来已经存在的依赖的基础上多添加了hbase-endpoint的依赖：

```
<dependency>
    <groupId>org.apache.hbase</groupId>
    <artifactId>hbase-endpoint</artifactId>
    <version>2.3.4</version>
</dependency>
```

代码7.3　HbaseAgg.java

```
01 package org.hadoop.hbase;
02 import org.apache.hadoop.conf.Configuration;
03 import org.apache.hadoop.hbase.CompareOperator;
04 import org.apache.hadoop.hbase.HBaseConfiguration;
05 import org.apache.hadoop.hbase.TableName;
06 import org.apache.hadoop.hbase.client.*;
07 import org.apache.hadoop.hbase.client.coprocessor.AggregationClient;
08 import org.apache.hadoop.hbase.client.coprocessor.LongColumnInterpreter;
09 import org.apache.hadoop.hbase.filter.SingleColumnValueFilter;
10 import org.apache.hadoop.hbase.util.Bytes;
11 import org.junit.After;
```

```java
12  import org.junit.Before;
13  import org.junit.Test;
14  public class HbaseAgg {
15      private Connection con;
16      private Configuration conf;
17      @Before
18      public void before() throws Exception {
19          conf = HBaseConfiguration.create();
20          conf.set("hbase.zookeeper.property.clientPort", "2181");
21          conf.set("hbase.zookeeper.quorum", "server201");
22          con = ConnectionFactory.createConnection(conf);
23      }
24      @After
25      public void after() throws Exception {
26          con.close();
27      }
28      /**
29       * 统计行数
30       */
31      @Test
32      public void testCount() throws Throwable {
33          if (con != null) {
34              AggregationClient aggregationClient = new AggregationClient(conf);
35              Scan scan = new Scan();
36              Long count = aggregationClient.rowCount(TableName.valueOf("ns1:stud"),
37                      new LongColumnInterpreter(), scan);
38              System.out.println("行数: " + count);
39              aggregationClient.close();
40              con.close();
41          } else {
42              System.out.println("连接失败");
43          }
44      }
45      /**
46       * 需要注意的是，必须写入xx字段为Long类型才可以进行统计，否则总是返回Null值
47       * 如果是在命令行，请使用stud.incr 'ROWKEY','f:xx',89的方式写入，直接就是Long类型
48       */
49      @Test
50      public void testMax() throws Throwable {
51          if (con != null) {
52              AggregationClient aggregationClient = new AggregationClient(conf);
53              Scan scan = new Scan();
54              scan.addColumn(Bytes.toBytes("f"), Bytes.toBytes("height"));
55              TableName table = TableName.valueOf("ns1:stud");
56              LongColumnInterpreter columnInterpreter = new LongColumnInterpreter();
57              Long max = aggregationClient.max(table, columnInterpreter, scan);
58              System.out.println("Max: " + max);
59              aggregationClient.close();
60              con.close();
61          } else {
62              System.out.println("连接失败");
63          }
64      }
65      /**
66       * 需要注意的是，必须写入xx字段为Long类型才可以进行统计，否则总是返回Null值
67       * 如果是在命令行，请使用stud.incr 'ROWKEY','f:age',89的方式写入，直接就是Long类型
```

```java
 68          */
 69         @Test
 70         public void testAvg() throws Throwable {
 71             if (con != null) {
 72                 AggregationClient aggregationClient = new AggregationClient(conf);
 73                 Scan scan = new Scan();
 74                 scan.addColumn(Bytes.toBytes("f"), Bytes.toBytes("height"));
 75                 TableName table = TableName.valueOf("ns1:stud");
 76                 LongColumnInterpreter columnInterpreter = new LongColumnInterpreter();
 77                 Double avg = aggregationClient.avg(table, columnInterpreter, scan);
 78                 System.out.println("avg is: " + avg);//avg is: 83.0
 79                 aggregationClient.close();
 80                 con.close();
 81             } else {
 82                 System.out.println("连接失败");
 83             }
 84         }
 85         /**
 86          * 添加查询条件
 87          * 添加一些条件，以下为f:name=xx的条件
 88          */
 89         @Test
 90         public void testSumWithFilter() throws Throwable {
 91             if (con != null) {
 92                 AggregationClient aggregationClient = new AggregationClient(conf);
 93                 Scan scan = new Scan();
 94                 scan.addColumn(Bytes.toBytes("f"), Bytes.toBytes("height"));
 95                 SingleColumnValueFilter filter =
 96                         new SingleColumnValueFilter(Bytes.toBytes("f"),
 97 Bytes.toBytes("name"), CompareOperator.EQUAL, Bytes.toBytes("Jack"));
 98                 scan.setFilter(filter);
 99                 TableName table = TableName.valueOf("ns1:stud");
100                 LongColumnInterpreter columnInterpreter = new LongColumnInterpreter();
101                 Double avg = aggregationClient.avg(table, columnInterpreter, scan);
102                 System.out.println("===>avg is: " + avg);//
103                 aggregationClient.close();
104                 con.close();
105             } else {
106                 System.out.println("连接失败..");
107             }
108         }
109 }
```

7.6 Phoenix

Phoenix是HBase的一个处理引擎，它让我们使用原生SQL就可以实现对HBase数据的CRUD，且Phoenix维护所有HBase的表，在创建时会自动创建协处理器。

Phoenix的官方网址为http://www.apache.org/dyn/closer.lua/phoenix。目前Phoenix 5.1版本支持HBase 2.3和HBase 2.4。以下是支持列表：

```
Phoenix Version    Release Date    Download
5.1.0    10/feb/2021
src    [ sha512 | asc ]
hbase-2.1-bin    [ sha512 | asc ]
hbase-2.2-bin    [ sha512 | asc ]
hbase-2.3-bin    [ sha512 | asc ]
hbase-2.4-bin    [ sha512 | asc ]
```

在安装Phoenix前，必须先停止HBase的所有进程。

步骤 01 下载Phoenix，需要对应HBase的版本。

下载地址：https://mirrors.bfsu.edu.cn/apache/phoenix/phoenix-5.1.0/phoenix-hbase-2.3-5.1.0-bin.tar.gz。

步骤 02 安装Phoenix。

解压：

```
$ tar -zxvf phoenix-hbase-2.3-5.1.0-bin.tar.gz -C /app/
```

修改hbase-site.xml文件。同时将hbase-site.xml文件复制到PHOENIX_HOME/bin目录下的hbase-site.xml文件中，并保持Phoenix中的配置与HBase中的hbase-site.xml文件配置一致。如果初次配置不成功，可以删除hdfs://server201:8020/hbase目录和/app/datas/zookeeper两个目录，然后再重新启动HBase。

完整的hbase-site.xml文件配置如下：

```xml
<configuration>
    <property>
        <name>hbase.tmp.dir</name>
        <value>/app/datas/hbase/tmp</value>
    </property>
    <property>
        <name>hbase.rootdir</name>
        <value>hdfs://server201:8020/hbase</value>
    </property>
    <property>
        <name>hbase.cluster.distributed</name>
        <value>true</value>
    </property>
    <property>
        <name>hbase.zookeeper.quorum</name>
        <value>server201:2181</value>
    </property>
    <!--以下三个为phoenix的配置-->
    <property>
        <name>phoenix.schema.isNamespaceMappingEnabled</name>
        <value>true</value>
    </property>
    <property>
        <name>hbase.regionserver.wal.codec</name>
        <value>
        org.apache.hadoop.hbase.regionserver.wal.IndexedWALEditCodec
        </value>
    </property>
    <property>
        <name>phoenix.schema.mapSystemTablesToNamespace</name>
        <value>true</value><!--此值默认就是true，可省略-->
    </property>
</configuration>
```

配置好以后，将此文件直接复制到phoenix/bin目录下：

```
$ cp /app/hbase-2.3.4/conf/hbase-site.xml /app/phoenix-5.1.0/bin/
```

复制server.jar文件。将phoenix-server-hbase-xxx.jar复制到HBase的lib目录下：

```
$ cp phoenix-server-hbase-2.3-5.1.0.jar /app/hbase-2.3.4/lib/
```

重新启动HBase：

```
$ hbase-2.3.4/bin/start-hbase.sh
```

然后登录Phoenix的SQL客户端：

```
[root@server201 bin]# ./sqlline.py server201:2181
Setting property: [incremental, false]
Setting property: [isolation, TRANSACTION_READ_COMMITTED]
0: jdbc:phoenix:server201:21>
```

步骤 03 使用Phoenix。

登录以后，输入!help查看所有的帮助命令：

```
0: jdbc:phoenix:server201:2181> !help
!all              Execute the specified SQL against all the current
                  connections
!appconfig        Set custom application configuration class name
!autocommit       Set autocommit mode on or off
!batch            Start or execute a batch of statements
!brief            Set verbose mode off
!call             Execute a callable statement
!close            Close the current connection to the database
!closeall         Close all current open connections
!columns          List all the columns for the specified table
!commandhandler   Add a command handler
...
```

使用!tables显示表：

```
0: jdbc:phoenix:server201:2181> !Tables
| TABLE_CAT | TABLE_SCHEM | TABLE_NAME | TABLE_TYPE
|           | SYSTEM      | CATALOG    | SYSTEM TABLE
...
```

使用!quit可以退出Phoenix。

Phoenix支持的数据类型如表7-5所示。

表7-5 Phoenix 支持的数据类型

数据类型	数据类型	数据类型
INTEGER Type	UNSIGNED_FLOAT Type	UNSIGNED_DATE Type
UNSIGNED_INT Type	DOUBLE Type	UNSIGNED_TIMESTAMP Type
BIGINT Type	UNSIGNED_DOUBLE Type	VARCHAR Type
UNSIGNED_LONG Type	DECIMAL Type	CHAR Type
TINYINT Type	BOOLEAN Type	BINARY Type
UNSIGNED_TINYINT Type	TIME Type	VARBINARY Type

(续表)

数据类型	数据类型	数据类型
SMALLINT Type	DATE Type	ARRAY
UNSIGNED_SMALLINT Type	TIMESTAMP Type	
FLOAT Type	UNSIGNED_TIME Type	

Phoenix支持的SQL操作可以参考官方网址：http://phoenix.apache.org/language/index.html。

下面来看看使用示例。

创建一个表：

```
create table books(id integer,name varchar(50),constraint pk primary key(id));
```

写入一行记录，注意使用的是upsert：

```
upsert into  books values(3,'Hadoop');
```

查询：

```
0: jdbc:phoenix:server31,server32,server33:21> select * from books;
+-----+--------+
| ID  | NAME   |
+-----+--------+
| 3   | Hadoop |
+-----+--------+
2 rows selected (0.042 seconds)
```

修改：

```
upsert into books values(1,'Phoenix');
```

删除：

```
delete from books where id=3;
```

更多操作不再赘述。通过上面的操作可以看出，Phoenix简化了HBase的查询语句。

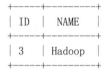 **JavaAPI**。

同样地，通过JDBC就可以快速通过Phoenix连接到HBase并执行操作，且可以使用读者熟悉的SQL语句。现在我们在项目中做以下工作：

（1）添加依赖。

（2）将hbase-site.xml文件放到项目的classpath目录下，即放到src/main/java/resources目录下。

在pom.xml文件中，添加以下两个依赖：

```xml
<dependency>
    <groupId>org.apache.phoenix</groupId>
    <artifactId>phoenix-core</artifactId>
    <version>5.1.0</version>
</dependency>
<dependency>
    <groupId>org.apache.phoenix</groupId>
    <artifactId>phoenix-hbase-compat-2.4.0</artifactId>
```

```
        <version>5.1.0</version>
    </dependency>
```

然后将hbase-site.xml放到classpath下，如下所示：

chapter09/src/main/resources/hbase-site.xml

连接代码如下：

代码 7.4 PhoenixOperation.java

```
01 Class.forName("org.apache.phoenix.jdbc.PhoenixDriver");
02 String url = "jdbc:phoenix:server201:2181";
03 Connection con = DriverManager.getConnection(url);
04 System.out.println(con);
05 con.close();
```

查询数据代码如下：

代码 7.5 PhoenixOperation.java

```
01 @Test
02 public void queryTest() throws Exception {
03     Statement st = connection.createStatement();
04     ResultSet rs = st.executeQuery("select * from books");
05     while (rs.next()) {
06         int id = rs.getInt("id");
07         String name = rs.getString("name");
08         System.out.println("id:" + id + ",name:" + name);
09     }
10     rs.close();
11     st.close();
12 }
```

同样地，还可以创建一个命名空间，并在命名空间中创建一个表进行如下操作。这些语法在Phoenix官方网站都可以找到。

```
0: jdbc:phoenix:server201:2181> create schema db1;
No rows affected (0.507 seconds)
```

然后就可以在这个命名空间中创建一个表，并进行相关操作：

```
0: jdbc:phoenix:server201:2181> create table db1.stud(id integer,name
varchar(100),constraint pk primary key(id));
No rows affected (1.278 seconds)
0: jdbc:phoenix:server201:2181> select * from db1.stud;
+----+------+
| ID | NAME |
+----+------+
+----+------+
No rows selected (0.033 seconds)
0: jdbc:phoenix:server201:2181> upsert into db1.stud values(1,'Jerry');
1 row affected (0.085 seconds)
0: jdbc:phoenix:server201:2181> upsert into db1.stud values(2,'Mary');
```

```
1 row affected (0.014 seconds)
0: jdbc:phoenix:server201:2181> select * from db1.stud;
+----+-------+
| ID | NAME  |
+----+-------+
| 1  | Jerry |
| 2  | Mary  |
+----+-------+
2 rows selected (0.031 seconds)
```

我们还可以进行事务操作，以下是测试代码。

代码 7.6　PhoenixOperation.java

```
01 @Test
02 public void testTx() throws Exception {
03     System.err.println("是否是自动提交："+connection.getAutoCommit());//默认值为false
04     if(connection.getAutoCommit()){
05         connection.setAutoCommit(false);
06     }
07     String sql = "upsert into db1.stud(id,name) values(3,'Alex')";
08     Statement st = connection.createStatement();
09     st.execute(sql);
10     connection.commit();//必须手动提交
11     connection.setAutoCommit(true);
12     st.close();
13     System.out.println("写入成功");
14 }
```

在Phoenix中使用聚合函数，直接在SQL中添加聚合函数就可以了，示例代码如下：

代码 7.7　PhoenixOperation.java

```
01 @Test
02 public void testAgg() throws Exception{
03     String sql = "select count(1) from db1.stud";
04     Statement st = connection.createStatement();
05     ResultSet rs = st.executeQuery(sql);
06     if (rs.next()) {
07         Long sum = rs.getLong(1);
08         System.out.println("count is:" + sum);
09     }
10     rs.close();
11     st.close();
12 }
```

最后，上例的所有连接都仅用于测试。在正式生产环境的连接中，请使用连接池维护多个连接，以减少客户端重复获取连接的时间。

7.7 小　　结

- HBase是指Hadoop DataBase，是面向列的数据库。
- HBase具有"大"的特点，一个表可以保存上亿级别的数据。
- HBase利用HDFS实现分布式的存储。
- HBase保存的数据，都是二进制形式，没有数据类型。
- HBase的主要进程是HMaster和HRegionServer。
- HBase伪分布配置时可以使用HBase内置的ZooKeeper，也可以使用外部独立的ZooKeeper，只要在hbase-env.sh中修改HBASE_MANAGER_ZK=false，且在hbase-site.xml中添加配置hbase.zookeeper.quorum=ip:2181即可。外部独立的ZooKeeper既可以是单节点，也可以是集群。
- Java代码连接HBase，无论是伪分布式或是真分布式，只要配置连接到ZooKeeper，即hbaseConfiguration.set("hbase.zookeeper.quorum","ip:2181");，即可以成功连接HBase。但在连接之前一定要配置HOSTS文件，指定连接的IP地址中主机名与IP地址的对应关系。

第 8 章

Flume数据采集实战

主要内容：

- Flume简介。
- Flume安装与配置。
- Flume部署。

Flume是Cloudera提供的一个高可用、高可靠、分布式的海量日志采集、聚合和传输系统。Flume支持定制各类数据源，如Avro、Thrift、Spooling等。同时，Flume提供对数据的简单处理，并将数据处理结果写入各种数据接收方，如将数据写入到HDFS文件系统中。

Flume作为Cloudera开发的实时日志收集系统，受到了业界的认可并得到广泛的应用。2010年11月，Cloudera开源了Flume的第一个可用版本0.9.2，这个系列版本被统称为Flume-OG（Original Generation）。随着Flume功能的扩展，Flume-OG代码开始臃肿、核心组件设计不合理、核心配置不标准等缺点暴露出来，尤其是在Flume-OG的最后一个发行版本0.94.0中，日志传输不稳定的现象尤为严重。为了解决这些问题，2011年10月Cloudera重构了核心组件、核心配置和代码架构，重构后的版本统称为Flume-NG（Next Generation）。改动的另一原因是将Flume纳入Apache旗下，Cloudera Flume改名为Apache Flume。

Flume的数据流由事件（Event）贯穿始终。事件是Flume的基本数据单位，它携带日志数据（字节数组形式）并且携带有头信息，这些Event由Agent外部的Source生成，当Source捕获事件后会进行特定的格式化，然后Source会把事件推入（单个或多个）Channel中。可以把Channel看作一个缓冲区，它将保存事件直到Sink处理完该事件。Sink负责持久化日志或者把事件推向另一个Source。

Flume以Agent为最小的独立运行单位。一个Agent就是一个JVM。Agent由Source、Channel和Sink三大组件构成，如图8-1所示。

值得注意的是，Flume提供了大量内置的Source、Channel和Sink。不同类型的Source、Channel和Sink可以自由组合。组合方式基于用户的配置文件，非常灵活。比如，Channel可以把事件暂存在内存里，也可以持久化到本地硬盘上。Sink可以把日志写入HDFS、HBase，甚至是另外一个Source等。Flume支持用户建立多级流，也就是多个Agent可以协同工作，并且支持Fan-in（扇入）、Fan-out（扇出）、Contextual Routing、Backup Routes，如图8-2所示。

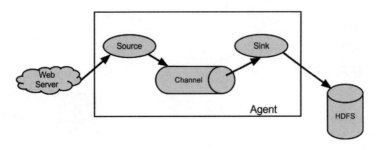

图8-1　Agent的组件构成

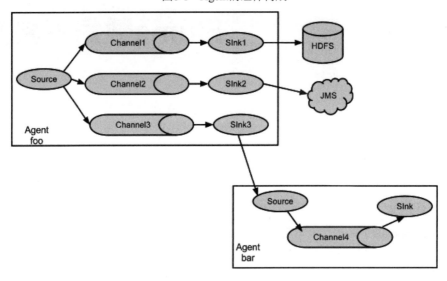

图8-2　Source、Channel和Sink的自由组合

Flume的一些核心概念如下：

- Agent：使用JVM运行Flume。每台机器运行一个Agent，但是可以在一个Agent中包含多个Source和Sink。
- Source：从Client收集数据，传递给Channel。
- Channel：连接Source和Sink，Channel缓存从Source收集来的数据。
- Sink：从Channel收集数据，并将数据写到目标文件系统中。

8.1　Flume的安装与配置

在安装Flume之前，需要确认已经安装了JDK，并正确配置了环境变量。

步骤01　下载并解压Flume。

下载地址：http://www.apache.org/dyn/closer.lua/flume/1.9.0/apache-flume-1.9.0-bin.tar.gz。

解压：

```
$ tar -zxvf ~/apache-flume-1.9.0-bin.tar.gz -C /app/
```

步骤 02 配置flume-env.sh文件。

在flume-env.sh文件中配置JAVA_HOME环境变量:

```
$ cp flume-env.sh.template flume-env.sh
$ vim flume-env.sh
export JAVA_HOME=/usr/local/java/jdk1.8.0_211
```

步骤 03 配置Flume的环境变量。

```
export FLUME_HOME=/home/isoft/app/flume-1.9.0
export PATH=$PATH:$FLUME_HOME/bin
```

使环境变量生效:

```
$ source ~/.bash_profile
```

现在可以使用version测试Flume的版本:

```
$ flume-ng version
flume 1.9.0
```

至此,Flume安装与配置已经完成。非常简单,以下将是部署两个基本的Flume Agent来测试Flume。

8.2 快速示例

根据官方网址示例,我们做一个快速的Flume示例,这个示例从网络上收集数据,并输出到日志里面。

步骤 01 定义配置文件。

在任意的目录下定义一个资源文件(如agent_1.conf),并输入以下内容。其中#开始的为注释,可以不用输入。

```
#定义三个核心组件
a1.sources = r1
a1.channels = c1
a1.sinks = k1
# 定义sources,即头1的类型及绑定的端口
a1.sources.r1.type = netcat
a1.sources.r1.bind = localhost
a1.sources.r1.port = 44444
# 定义输出的目标为系统日志
a1.sinks.k1.type = logger
# 定义Channel的类型及大小
a1.channels.c1.type = memory
a1.channels.c1.capacity = 1000
a1.channels.c1.transactionCapacity = 100
# 将三个对象的关系进行整合
a1.sources.r1.channels = c1
# 注意以下Channel后面没有s
a1.sinks.k1.channel = c1
```

步骤02 启动Agent。

使用flume-ng命令，即可以启动一个Agent，通过--conf-files可以指定我们前面配置的配置文件。

```
$ /app/flume-1.9.0/bin/flume-ng agent --conf /home/isoft/app/flume-1.9.0/conf
 --conf-file /app/conf/flume/agent_1.conf \
 --name a1 -Dflume.root.logger=INFO,console
```

说明：

- --conf：用于指定Flume配置文件的目录，这个目录下包含flume-env.sh等文件。
- --conf-file：用于指定用户自己配置的Agent文件。
- --name：用于指定用户配置文件中的Agent名称。
- -Dxx：用于配置系统一些变量。比如配置日志输出格式，flume.root.logger=INFO,console，注意console的c是小写的。

步骤03 访问44444端口并输入数据。

```
[hadoop@server201 ~]$ telnet localhost 44444
Trying ::1...
telnet: connect to address ::1: Connection refused
Trying 127.0.0.1...
Connected to localhost.
Escape character is '^]'.
Jack
Mary
Rose
```

查看flume-ng的控制台日志：

```
{ headers : {} body : A4 61 63 6B 0D      Jack. }
{ headers : {} body : A4 61 63 6B 0D      Mary. }
{ headers : {} body : A4 61 63 6B 0D      Rose. }
```

至此一个简单的Flume Agent就配置完成了。

8.3　在 ZooKeeper 中保存 Flume 的配置文件

我们也可以将配置文件存入ZooKeeper中。需要做以下的功能：

（1）启动Agent时，使用-z指定ZooKeeper的主机地址。
（2）使用-p指定ZooKeeper的节点名称。
（3）将Flume的配置信息，保存到ZooKeeper的节点数据中。
（4）在$FLUME_HOME目录下，创建plugins.d目录，并创建ZooKeeper子目录，用于保存Flume依赖的ZooKeeper的jar包。
（5）ZooKeeper节点保存的数据最多为1MB。

根据官方文档的说明，Flume在ZooKeeper节点的结构类似以下结果：

```
- /flume
  |- /a1 [Agent config file]
  |- /a2 [Agent config file]
```

启动时，通过指定-z、-p两个参数，使用ZooKeeper中的节点信息：

```
$ bin/flume-ng agent -conf conf -z zkhost:2181,zkhost1:2181 -p /flume -name a1
-Dflume.root.logger=INFO,console
```

两上参数的具体含义如下：

- z：ZooKeeper连接字符串，冒号分隔的"主机名:端口"列表。
- p：ZooKeeper中的Flume Base Path，用于存储代理配置。

为了可以将Flume的配置文件写入到ZooKeeper节点的data中，我们需要编写一个Java类，然后在Linux上使用java命令将指定的文件上传到ZooKeeper的节点数据中去。

步骤01 开发ZooKeeper上传文件的程序。

添加依赖：

```xml
<dependency>
    <groupId>org.apache.zookeeper</groupId>
    <artifactId>zookeeper</artifactId>
    <version>3.6.2</version>
</dependency>
```

代码8.1 ZkUpload.java

```
01 package org.hadoop.flume;
02 import org.apache.zookeeper.*;
03 import org.apache.zookeeper.data.Stat;
04 import java.io.ByteArrayOutputStream;
05 import java.io.FileInputStream;
06 import java.io.InputStream;
07 import java.util.concurrent.CountDownLatch;
08 public class ZkUpload {
09     /**
10      * 上传文件
11      */
12     public static void main(String[] args) throws Exception {
13         if (args.length < 3) {
14             System.err.println("用法: <zkServer> <path> <file>");
15             return;
16         }
17         String zkHosts = args[0];
18         String path = args[1];
19         String file = args[2];
20         CountDownLatch countDownLatch = new CountDownLatch(1);
21         ZooKeeper zooKeeper = new ZooKeeper(zkHosts, 3000, new Watcher() {
22             @Override
23             public void process(WatchedEvent event) {
24                 System.out.println("连接" + zkHosts + "成功");
25                 countDownLatch.countDown();
26             }
27         });
28         countDownLatch.await();
29         try {
```

```
30              //检查文件是否存在
31              Stat stat = zooKeeper.exists(path, null);
32              if (stat != null) {
33                  throw new RuntimeException("节点已经存在！");
34              }
35              //开始上传文件
36              InputStream in = new FileInputStream(file);
37              //判断文件大小，最大是1MB
38              int size = in.available();
39              if (size > (1024 * 1024)) {
40                  throw new RuntimeException("上传的文件大小不能超过1M");
41              }
42              ByteArrayOutputStream bout = new ByteArrayOutputStream();
43              byte[] bs = new byte[1024 * 4];
44              int len = 0;
45              while ((len = in.read(bs)) != -1) {
46                  bout.write(bs, 0, len);
47              }
48              bout.close();
49              bs = bout.toByteArray();
50              path = zooKeeper.create(path, bs, ZooDefs.Ids.OPEN_ACL_UNSAFE,
    CreateMode.PERSISTENT);
51              System.out.println("节点创建成功: " + path);
52          } finally {
53              zooKeeper.close();
54          }
55      }
56 }
```

步骤02 将代码上传到Linux服务器并编译。

上传代码，并直接使用javac编译ZkUpload.java：

```
$ javac -classpath .:/app/zookeeper-3.6.2/lib/* -d . ZkUpload.java
```

运行测试：

```
[hadoop@server201 java]$ java org.hadoop.flume.ZkUpload
用法：<zkServer> <path> <file>
```

如果显示了用法，则说明代码可用。

步骤03 启动ZooKeeper并上传配置文件。

```
$ java -jar zk.jar org.hadoop.flume.ZkUpload  localhost:2181  /flume/a1
/app/datas/conf/flume/agent_1.conf
```

查看ZooKeeper上/flume/a1目录里面的内容：

```
[zk: localhost:2181(CONNECTED) 3] ls /flume
[a1]
```

以下查看内容，可见已经上传到节点的数据中：

```
[zk: localhost:2181(CONNECTED) 4] get /flume/a1
# 定义三个核心组件
a1.sources = r1
a1.channels = c1
a1.sinks = k1
```

```
# 定义sources即头1的类型及绑定的端口
a1.sources.r1.type = netcat
a1.sources.r1.bind = localhost
a1.sources.r1.port = 44444
# 定义输出的目标为系统日志
a1.sinks.k1.type = logger
# 定义channel的类型及大小
a1.channels.c1.type = memory
a1.channels.c1.capacity = 1000
a1.channels.c1.transactionCapacity = 100
# 将三个对象的关系进行整合
a1.sources.r1.channels = c1
# 注意以下channel后面没有s
a1.sinks.k1.channel = c1
```

步骤04 给Flume添加ZooKeeper依赖。

由于启动flume-agent时需要读取ZooKeeper中的数据,此时需要依赖ZooKeeper。根据Flume官方提示,可以在$FLUME_HOME目录下创建plugins.d目录,并将依赖保存到这个目录下,plugins.d的目录结构应该为:

```
plugins.d/
plugins.d/custom-source-1/
plugins.d/custom-source-1/lib/my-source.jar
plugins.d/custom-source-1/libext/spring-core-2.5.6.jar
plugins.d/custom-source-2/
plugins.d/custom-source-2/lib/custom.jar
plugins.d/custom-source-2/native/gettext.so
```

目录结构说明:

Each plugin (subdirectory) within plugins.d can have up to three sub-directories:

(1) lib: 扩展jar文件。

(2) libext: 依赖的扩展jar文件。

(3) native: 本地库,如c语言编译文件.so。

现在我们将ZooKeeper的所有jar包放到flume_home/plugins.d/zookeeper/lib目录下:

```
$ ln -s /home/isoft/app/zookeeper-3.5.5/lib/* \
    /home/isoft/app/flume-1.9.0/plugins.d/zookeeper/lib
```

步骤05 现在启动flume-agent。

```
$ flume-ng agent --conf /app/app/flume-1.9.0/conf/ \
-z server201:2181 \
-p /flume -name a1 \
-Dflume.root.logger=INFO,console
```

启动成功,将显示以下信息:

```
Starting Channel c1
Starting Sink k1
Starting Source r1
Created serverSocket:sun.nio.ch.ServerSocketChannelImpl[/127.0.0.1:44444
```

到此,就可以使用ZooKeeper集群中配置的文件了。

8.4 Flume 的更多 Source

所有Flume的Source都可以在官方网址上找到它的配置信息。

8.4.1 Avro Source

Avro Source可以定制avro-client发送一个指定的文件给Flume Agent，Avro源使用Avro RPC机制。Flume主要的RPC Source也是Avro Source，它使用Netty-Avro inter-process的通信（IPC）协议来通信，因此可以用Java或JVM语言发送数据到Avro Source端。它的配置文件主要包含以下三个参数：

- type：Avro source的别名是avro，也可以使用完整类名称org.apache.flume.source.AvroSource。
- bind：绑定的IP地址或主机名。使用0.0.0.0绑定机器所有端口。
- port：绑定监听端口。

1. 通过avro-client发送文件数据

flume-ng自带avro-client命令，可以将指定的文件通过avro rpc发送到指定的Avro源。

当输入flume-ng help命令后，就可以查看到这个avro-client命令：

```
[hadoop@server201 flume-1.9.0]$ bin/flume-ng help
Usage: bin/flume-ng <command> [options]...
commands:
  help                    display this help text
  agent                   run a Flume agent
  avro-client             run an avro Flume client
  version                 show Flume version info
```

Avro Source配置示例：

```
#配置三个源
a1.sources=r1
a1.channels=c1
a1.sinks=k1
#配置r1的输入源为avro
a1.sources.r1.type=avro
a1.sources.r1.bind=0.0.0.0
a1.sources.r1.port=4141
#配置channels的类型为memory
a1.channels.c1.type=memory
a1.channels.c1.capacity=1000
a1.channels.c1.transactionCapacity=100
#配置sinks的目标为logger
a1.sinks.k1.type=logger
#组织三个组件
a1.sources.r1.channels=c1
a1.sinks.k1.channel=c1
```

现在启动Avro Source：

```
$ flume-ng agent -n a1 -c conf/ --conf-file conf/01_avro.conf \
```

```
-Dflume.root.logger=INFO,consol
```

然后使用flume-ng avro-client命令向4141端口发送文件数据：

```
$flume-ng avro-client -c conf/ -H 127.0.0.1 -p 4141 -F hello.txt
```

现在就可以查看日志log/flume.log中的内容：

```
{ headers : {} body : .....}
{ headers : {} body : .....}
{ headers : {} body : .....}
```

2. Java API向avro源发送数据

同样，也可以使用Java代码向Avro源发送数据。以下是Flume Agent的配置参考示例：

```
a1.channels = c1
a1.sources = r1
a1.sinks = k1
a1.channels.c1.type = memory
a1.sources.r1.channels = c1
a1.sources.r1.type = avro
# For using a thrift source set the following instead of the above line.
# a1.source.r1.type = thrift
a1.sources.r1.bind = 0.0.0.0
a1.sources.r1.port = 41414
a1.sinks.k1.channel = c1
a1.sinks.k1.type = logger
```

使用上面的配置文件，并启动Agent。

首先，需要添加依赖：

```xml
<dependency>
    <groupId>org.apache.flume</groupId>
    <artifactId>flume-ng-core</artifactId>
    <version>1.9.0</version>
</dependency>
```

开发一个基本的测试功能，此测试向Flume监听的端口发送数据。如代码8.2所示。

代码8.2 SendData.java

```
01 package org.hadoop.flume;
02 import org.apache.flume.Event;
03 import org.apache.flume.api.RpcClient;
04 import org.apache.flume.api.RpcClientFactory;
05 import org.apache.flume.event.EventBuilder;
06 import org.junit.Test;
07 import java.nio.charset.Charset;
08 public class SendData {
09     @Test
10     public void sendData() throws Exception {
11         String host = "server201";
12         int port = 4141;
13         RpcClient client = RpcClientFactory.getDefaultInstance(host, port);
14         //发送数据
15         Event event = EventBuilder.withBody("Hello", Charset.forName("UTF-8"));
16         client.append(event);
```

```
17        client.close();//关闭
18     }
19 }
```

上面的代码中，通过client.append将Event发送到Avro Source源上去。

连续发送Event的示例如代码8.3所示。

代码 8.3　SendData2.java

```
01 package org.hadoop.flume;
02 import org.apache.flume.Event;
03 import org.apache.flume.api.RpcClient;
04 import org.apache.flume.api.RpcClientFactory;
05 import org.apache.flume.event.EventBuilder;
06 import java.nio.charset.Charset;
07 import java.util.Scanner;
08 public class SendData2 {
09     public static void main(String[] args) throws Exception {
10         String host = "server201";
11         int port = 4141;
12         RpcClient client = RpcClientFactory.getDefaultInstance(host, port);
13         //连续输入数据
14         Scanner sc = new Scanner(System.in);
15         while (true) {
16             String str = sc.nextLine();
17             if (str.equals("exit")) {
18                 break;
19             }
20             Event event = EventBuilder.withBody(str, Charset.forName("UTF-8"));
21             client.append(event);
22         }
23         client.close();
24     }
25 }
```

启动Java程序并连续输入数据，然后查看Flume收到的数据，结果如下：

```
{ header : {} bydy : A1 B4 64 A3    Jack}
{ header : {} bydy : A1 B4 64 A3    Jack}
...
```

源代码说明：

在Flume 1.4以后，连接Avro Source或是连接Thrift Source都可以使用RpcClient。RpcClient的两个主要子类为：ThriftRpcClient和NettyAvroRpcClient，如图8-3所示。

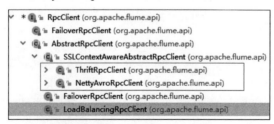

图8-3　RpcClient的两个主要子类

上面的代码，由于连接是Avro Source源，所以返回的对象为NettyAvroRpcClient，即代码：

```
RpcClient client = RpcClientFactory.getDefaultInstance(host,port);
```

结果返回的是NettyAvroRpcClient。

3. 开发一个netcat源向Avro发送数据

下面开发一个示例配置，将配置两个Agent，两个Agent中间通过Avro RPC进行通信，配置图示如图8-4所示。

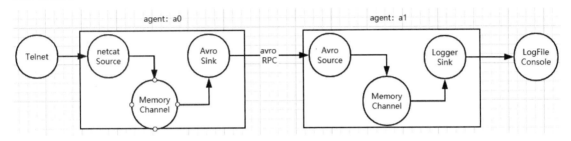

图8-4　Agent配置图示

修改avro.conf内容如下：

```
#配置agent a0
a0.sources=r0
a0.channels=c0
a0.sinks=k0
a0.sources.r0.type=netcat
a0.sources.r0.bind=0.0.0.0
a0.sources.r0.port=4040
a0.channels.c0.type=memory
a0.channels.c0.capacity=1000
a0.channels.c0.transactionCapacity=100
#配置sink为avro并向一个固定的ip和端口号4141
a0.sinks.k0.type=avro
a0.sinks.k0.hostname=127.0.0.1
a0.sinks.k0.port=4141
a0.sources.r0.channels=c0
a0.sinks.k0.channel=c0
##############################
#配置agent a1，内容同上直接复制
a1.sources=r1
a1.channels=c1
a1.sinks=k1
#配置source为avro，监听本机的4141端口
a1.sources.r1.type=avro
a1.sources.r1.bind=0.0.0.0
a1.sources.r1.port=4141
a1.channels.c1.type=memory
a1.channels.c1.capacity=1000
a1.channels.c1.transactionCapacity=100
a1.sinks.k1.type=logger
a1.sources.r1.channels=c1
a1.sinks.k1.channel=c1
```

注意，上面的4141就是指上面的Avro接收数据的端口。

现在启动Agent a1：

```
$ bin/flume-ng agent -n a1 -c conf/ --conf-file conf/avro.conf \
-Dflume.root.logger=INFO,console
```

然后再启动Agent a0：

```
$bin/flume-ng agent -n a0 -c conf/ --conf-file conf/avro.conf \
-Dflume.root.logger=INFO,console
```

现在就可以使用telnet登录，并向这个netcat发送数据了，然后这个数据会通过Avro RPC发送到Avro源，最后输出到日志文件中。

以下是向Agent a0发送数据：

```
[hadoop@server201 ~]$ telnet localhost 4040
Trying ::1...
Connected to localhost.
Escape character is '^]'.
Jack
OK
```

以下是Agent a1输出到日志的数据：

```
{ headers:{} body: 4A 61 63 6B 0D                         Jack. }
{ headers:{} body: 4A 61 63 6B 0D                         Jack. }
{ headers:{} body: 4A 61 63 6B 0D                         Jack. }
```

至此，两个Agent通过Avro RPC通信的示例演示成功了。

8.4.2 Thrift Source 和 Thrift Sink

Thrift与Avro同样可以进行串联。Thrift与Avro类似，都可以实现RPC调用，但Thrift支持更多的语言。

1. 使用Java代码向Thrift发送数据

与Avro类似，也是通过RpcClient向Thrift Source发送数据。

首先定义一个Agent 配置文件thrift.conf，内容如下：

```
#定义三个组件：
a1.sources=r1
a1.channels=c1
a1.sinks=k1
#定义Thrift Source，绑定本机所有ip的4141端口
a1.sources.r1.type=thrift
a1.sources.r1.bind=0.0.0.0
a1.sources.r1.port=4141
#channel依然使用内存
a1.channels.c1.type=memory
a1.channels.c1.capacity=1000
a1.channels.c1.transactionCapacity=100
#sink为日志
a1.sinks.k1.type=logger
```

```
#绑定三个组件的关系
a1.sources.r1.channels=c1
a1.sinks.k1.channel=c1
```

现在启动Agent a1:

```
$bin/flume-ng agent -n a1 -c conf/ --conf-file conf/thrift.conf \
-Dflume.root.logger=INFO,console
```

现在开发Java客户端,用于发送数据。

通过RpcClientFactory.getThriftInstance(host,port)方法,可以获取一个Thrift的RpcClient对象,如代码8.4所示。

代码8.4　ThriftClient.java

```
01 package org.hadoop.flume;
02 import org.apache.flume.Event;
03 import org.apache.flume.api.RpcClient;
04 import org.apache.flume.api.RpcClientFactory;
05 import org.apache.flume.event.EventBuilder;
06 import org.junit.Test;
07 import java.nio.charset.Charset;
08 public class ThriftClient {
09     @Test
10     public void test1() throws Exception {
11         String host = "server101";
12         int port = 4141;
13         RpcClient client = RpcClientFactory.getThriftInstance(host, port);
14         System.err.println(client);
15         Event event = EventBuilder.withBody("Hello", Charset.forName("UTF-8"));
16         client.append(event);
17         client.close();//关闭
18     }
19 }
```

现在就可以开发一个连续发送数据的测试,如代码8.5所示。

代码8.5　ThriftSender.java

```
01 package org.hadoop.flume;
02 import org.apache.flume.Event;
03 import org.apache.flume.api.RpcClient;
04 import org.apache.flume.api.RpcClientFactory;
05 import org.apache.flume.event.EventBuilder;
06 import java.nio.charset.Charset;
07 import java.util.Scanner;
08 public class ThriftSender {
09     public static void main(String[] args) throws Exception {
10         String host = "server201";
11         int port = 4141;
12         RpcClient client = RpcClientFactory.getThriftInstance(host,port);
13         //连续输入数据
14         Scanner sc = new Scanner(System.in);
15         while(true){
16             String str = sc.nextLine();
```

```
17              if(str.equals("exit")){
18                  break;
19              }
20              Event event = EventBuilder.withBody(str, Charset.forName("UTF-8"));
21              client.append(event);
22          }
23          client.close();
24      }
25 }
```

运行代码8.5,并输入一些数据,查看Flume日志接收到的数据:

```
{ headers:{} body: 48 65 6C 6C 6F             Hello }
{ headers:{} body: 48 65 6C 6C 6F             First }
{ headers:{} body: 48 65 6C 6C 6F             Second }
...
```

2. Thrift Sink

类似于Avro,我们也可以将多个Thrift进行串联,即Thrift Sink的输出为Thrift Source的输入。配置图如图8-5所示。

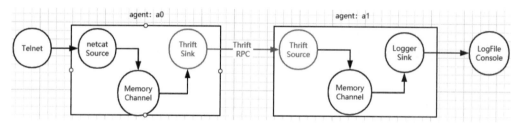

图8-5　Thrift Sink的配置图

配置文件thrift.conf:

```
#配置agent a0:
a0.sources=r0
a0.channels=c0
a0.sinks=k0
a0.sources.r0.type=netcat
a0.sources.r0.bind=0.0.0.0
a0.sources.r0.port=4040
a0.channels.c0.type=memory
a0.channels.c0.capacity=1000
a0.channels.c0.transactionCapacity=100
#定义thrift的输出
a0.sinks.k0.type=thrift
a0.sinks.k0.hostname=127.0.0.1
a0.sinks.k0.port=4141
#定义agent a0关联的三个组件
a0.sources.r0.channels=c0
a0.sinks.k0.channel=c0
############定义agent a1,即第二个agent
a1.sources=r1
a1.channels=c1
a1.sinks=k1
```

```
#定义thrift的souce即输入，注意这个端口号与上面thrift sink的端口号相同
a1.sources.r1.type=thrift
a1.sources.r1.bind=0.0.0.0
a1.sources.r1.port=4141
a1.channels.c1.type=memory
a1.channels.c1.capacity=1000
a1.channels.c1.transactionCapacity=100
a1.sinks.k1.type=logger
#关联agent a1的三个组件
a1.sources.r1.channels=c1
a1.sinks.k1.channel=c1
```

启动Agent a1：

```
$ flume-ng agent -n a1 -c conf/ --conf-file conf/thrift.conf \
-Dflume.root.logger=INFO,console
```

启动Agent a0：

```
$ flume-ng agent -n a0 -c conf/ --conf-file conf/thrift.conf \
-Dflume.root.logger=INFO,console
```

现在通过telnet访问Agent a0的4040端口：

```
$ telnet 127.0.0.1 4040
Jack
ok
Mary
ok
```

现在，查看Flume Agent a1输出的日志信息：

```
{ headers:{} body: 4A 61 63 6B 0D                          Jack.}
{ headers:{} body: 4A 61 63 6B 0D                          Mary.}
```

8.4.3 Exec Source

Exec Source可以用来执行Linux命令，如执行tail -f somefile.log。

定义exec.conf配置文件，内容如下：

```
#定义三个组件
a1.sources=r1
a1.channels=c1
a1.sinks=k1
#定义Exec Source，执行的命令通过command定义
a1.sources.r1.type=exec
a1.sources.r1.command=tail -F /home/isoft/a.log
a1.channels.c1.type=memory
a1.channels.c1.capacity=1000
a1.channels.c1.transactionCapacity=100
a1.sinks.k1.type=logger
#组合三个组件
a1.sources.r1.channels=c1
a1.sinks.k1.channel=c1
```

启动Flume Agent：

```
$ bin/flume-ng agent -n a1 -c conf/ --conf-file conf/exec.conf \
```

```
-Dflume.root.logger=INFO,console
```

现在向a.log日志文件中追加数据：

```
$echo a >> a.log
$echo Jack >> a.log
$echo Mary >> a.log
```

查看日志的输出：

```
{ headers:{} body: 61                                          a }
{ headers:{} body: 61                                       Jack }
{ headers:{} body: 61                                       Mary }
```

8.4.4 Spool Source

Spool Source用于监控一个目录下文件的增加，主要是文本文件。

创建Flume Agent配置文件spool.conf，内容如下：

```
#声明三个组件
a1.sources=r1
a1.channels=c1
a1.sinks=k1
#定义Spool Source，读取指定的目录
a1.sources.r1.type=spooldir
a1.sources.r1.spoolDir=/home/isoft/a
a1.sources.r1.fileHeader=true
a1.channels.c1.type=memory
a1.channels.c1.capacity=1000
a1.channels.c1.transactionCapacity=100
a1.sinks.k1.type=logger
#组合这三个组件
a1.sources.r1.channels=c1
a1.sinks.k1.channel=c1
```

启动这个Agent：

```
$ flume-ng agent -n a1 -c conf/ --conf-file conf/spool.conf \
-Dflume.root.logger=INFO,console
```

将任意一个文本文件直接复制到/home/isoft/a目录下：

```
$ cp a.txt /home/hadoop/a/
```

然后查看Flume Agent输出的日志，由于设置了fileHeader=true，所以会携带header信息：

```
{ headers:{file=/home/isoft/a/a.log} body: 61                  a }
{ headers:{file=/home/isoft/a/a.log} body: 61 62 63          abc }
{ headers:{file=/home/isoft/a/a.log} body: 61 62 63         Jack }
{ headers:{file=/home/isoft/a/a.log} body: 61 62 63         Mike }
{ headers:{file=/home/isoft/a/a.log} body: 61 62 63         Rose }
```

8.4.5 HDFS Sinks

定义配置文件hdfs.conf，内容如下：

```
#定义三个组件
a1.sources=r1
```

```
a1.channels=c1
a1.sinks=k1
a1.sources.r1.type=netcat
a1.sources.r1.bind=0.0.0.0
a1.sources.r1.port=4141
a1.channels.c1.type=memory
a1.channels.c1.capacity=1000
a1.channels.c1.transactionCapacity=100
#定义HDFS Sink
a1.sinks.k1.type=hdfs
#输出的HDFS目录，/flume/events目录必须已经存在
a1.sinks.k1.hdfs.path=/flume/events/%Y%m%d/%H%M/%S
a1.sinks.k1.hdfs.filePrefix=FlumeData
#每10分钟一个子目录
a1.sinks.k1.hdfs.round=true
a1.sinks.k1.hdfs.roundUnit=minute
a1.sinks.k1.hdfs.roundValue=10
a1.sinks.k1.hdfs.useLocalTimeStamp=true  #必须设置为true才可使用%Y
a1.sinks.k1.hdfs.writeFormat=Text   #以文本形式输出，否则为字节码
#组合三个组件
a1.sources.r1.channels=c1
a1.sinks.k1.channel=c1
```

启动这个Agent：

```
$ flume-ng agent -n a1 -c conf/ --conf-file conf/hdfs.conf \
-Dflume.root.logger=INFO,console
```

然后通过telnet登录：

```
[hadoop@server201 ~]$ telnet 127.0.0.1 4141
Trying 127.0.0.1...
Connected to 127.0.0.1.
Escape character is '^]'.
Smith
Rose
```

通过hdfs dfs -text命令查看内容：

```
[hadoop@server201 ~]$ hdfs dfs -text /flume/events/2020310/0440/00/*
1570696963767    Smith
1570697055723    Rose
```

可见，内容已经添加到HDFS文件中去了。

8.5 小　　结

- Flume是数据采集工具，一般主要用于采集日志信息。
- Agent是Flume的核心，每一个Agent都包含Source、Channel、Sink。
- 多个Agent可以组成一个链。
- Flume定义了丰富的组件，只需要在配置文件中配置并组合它们即可。

第 9 章

Kafka实战

主要内容：

- ❖ Topic发布订阅。
- ❖ Broker。
- ❖ Partition。
- ❖ Kafka集群安装。

　　Kafka是一个分布式、可分区、多副本、多订阅者、基于ZooKeeper协调的分布式日志系统，也可以当作MQ消息队列系统，其发布订阅模式如图9-1所示。Kafka常用于Web/Nginx日志、访问日志、消息服务等场景，Linkedin于2010年贡献给了Apache基金会并成为顶级开源项目。

　　Kafka的主要应用场景是日志收集系统和消息系统。Kafka主要设计目标如下：

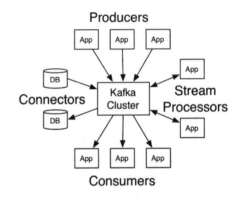

图9-1　发布订阅模式

- 以时间复杂度为O(1)的方式提供消息持久化能力，即使对TB级以上数据也能保证常数时间的访问性能。
- 高吞吐率，即使在非常廉价的商用机器上也能做到单机支持每秒100KB条消息的传输。
- 支持Kafka Server间的消息分区及分布式消费，同时保证每个分区（Partition）内的消息顺序传输。
- 同时支持离线数据处理和实时数据处理。
- Scale out，即支持在线水平扩展。

　　下载Kafka需要注意与当前安装的Scala版本相对应。本书选择kafka_2.12-2.7.0.tgz，其中2.12.0是指Scala的版本，2.7是Kafka的版本。其下载地址为http://kafka.apache.org。

9.1 Kafka 的特点

1. 解耦

在项目启动之初来预测将来项目会碰到什么需求,这是极其困难的。因此,消息系统在处理过程中间插入了一个隐含的、基于数据的接口层,两边的处理过程都要实现这一接口。这允许独立地扩展或修改两边的处理过程,只要确保它们遵守同样的接口约束。

2. 冗余(副本)

有些情况下,处理数据的过程会失败。除非数据被持久化,否则将造成丢失。消息队列把数据进行持久化,直到它们被完全处理,通过这一方式规避了数据丢失的风险。许多消息队列所采用的"插入－获取－删除"范式中,在把一个消息从队列中删除之前,需要我们的处理系统明确地指出该消息已经被处理完毕,从而确保数据被安全保存直到使用完毕。

3. 扩展性

因为消息队列解耦了处理过程,所以增大消息入队和处理的频率非常容易,只要另外增加处理过程即可。不需要改变代码、不需要调节参数。扩展就像调大音响的音量一样简单。

4. 灵活性与峰值处理能力

在访问量剧增的情况下,应用仍然需要继续发挥作用,但是这样的突发流量并不常见;如果为以能处理这类峰值访问为标准来投入随时待命的资源,无疑是巨大的浪费。使用消息队列能够使关键组件顶住突发的访问压力,而不会因为突发的超负荷请求而完全崩溃。

5. 可恢复性

系统的一部分组件失效时,不会影响到整个系统。消息队列降低了进程间的耦合度,所以即使一个处理消息的进程挂掉,加入队列中的消息仍然可以在系统恢复后被处理。

6. 顺序保证

在大多使用场景下,数据处理的顺序都很重要。大部分消息队列本来就是排序的,并且能保证数据按照特定的顺序来处理。Kafka保证一个Partition内的消息的有序性。

7. 缓冲

在任何重要的系统中,都会存在需要不同的处理时间的元素。例如,加载一幅图片比应用过滤器花费更少的时间。消息队列通过一个缓冲层来帮助任务以最高效率执行——写入队列的处理会尽可能地快速。该缓冲有助于控制和优化数据流经过系统的速度。

8. 异步通信

很多时候,用户不想也不需要立即处理消息。消息队列提供了异步处理机制,允许用户把一个消息放入队列，但并不立即处理它。想向队列中放入多少消息就放多少，然后在需要的时候再去处理它们。

9.2　Kafka 术语

1. Broker

　　Kafka集群包含一个或多个服务器，服务器节点称为Broker。Broker存储Topic的数据。如果某Topic有N个Partition，集群有N个Broker，那么每个Broker存储该Topic的一个Partition。

　　如果某Topic有N个Partition，集群有(N+M)个Broker，那么其中有N个Broker存储该Topic的一个Partition，剩下的M个Broker不存储该Topic的Partition数据。

　　如果某Topic有N个Partition，集群中Broker数目少于N个，那么一个Broker存储该Topic的一个或多个Partition。在实际生产环境中，尽量避免这种情况的发生，这种情况容易导致Kafka集群数据不均衡。

2. Topic

　　每条发布到Kafka集群的消息都有一个类别，这个类别被称为Topic（物理上不同Topic的消息分开存储，逻辑上一个Topic的消息虽然保存于一个或多个Broker上，但用户只需指定消息的Topic，即可生产或消费数据而不必关心数据存于何处）。Topic类似于数据库的表名。

3. Partition

　　Topic中的数据分割为一个或多个Partition，也称为分区。Topic是逻辑概念，Partition是物理概念，生产者只需要关心数据发送给哪一个Topic，消费者也只关心自己读取了哪一个Topic。每个Topic至少有一个Partition。每个Partition中的数据使用多个segment文件存储。Partition中的数据是有序的，不同Partition间的数据则不保证有序。如果Topic有多个Partition，消费数据时就不能保证数据的顺序。在需要严格保证消息的消费顺序的场景下，需要将Partition数目设置为1。

4. Producer

　　Producer（生产者）即数据的发布者，该角色将消息发布到Kafka的Topic中。Broker接收到生产者发送的消息后，Broker将该消息追加到当前用于追加数据的segment文件中。生产者发送的消息，存储到一个Partition中，生产者也可以指定数据存储的Partition。

5. Consumer

　　Consumer（消费者）可以从Broker中读取数据，消费者可以消费多个Topic中的数据。

6. Consumer Group

　　每个Consumer属于一个特定的Consumer Group（可为每个Consumer指定group name，若不指定group name，则属于默认的group）。

7. Leader

　　每个Partition有多个副本，其中有且仅有一个作为Leader，Leader是当前负责数据读写的Partition。

8. Follower

Follower（从服务）跟随Leader，所有写请求都通过Leader路由，数据变更会广播给所有Follower，Follower与Leader保持数据同步。如果Leader失效，则从Follower中选举出一个新的Leader。当Follower与Leader挂掉、卡住或者同步太慢，Leader会把这个Follower从ISR（In Sync Replicas）列表中删除，重新创建一个Follower。

9.3 Kafka 安装与部署

Kafka可以在一台机器上安装多个，只要在启动时指定不同的server.properties配置文件即可，此文件中配置的为主机名和端口。

9.3.1 单机部署

单机部署，指仅在单一的主机上安装Kafka，需要ZooKeeper的支持。

步骤01 安装ZooKeeper。

请参考之前章节中有关ZooKeeper安装与配置的相关内容。

配置zoo.properties：

```
tickTime=2000
initLimit=10
syncLimit=5
dataDir=/app/datas/zookeeper
clientPort=2181
# ZooKeeper与spark端口相同，需要修改，此值默认为8080
admin.serverPort=9999
```

配置环境变量（可选）：

```
export ZOOKEEPER_HOME=/app/zookeeper-3.6.2/
export PATH=$PATH:$ZOOKEEPER_HOME/bin
```

启动ZooKeeper：

```
[hadoop@server201 /app]$/app/zookeeper-3.6.2/bin/zkServer.sh start
ZooKeeper JMX enabled by default
Using config: /app/zookeeper-3.6.2/bin/../conf/zoo.cfg
Mode: standalone
```

步骤02 安装Kafka。

解压：

```
$ tar -zxvf kafka-2.12.0_2.7.0.tgz -C /app/
```

配置server.properties文件，因为ZooKeeper也是单机安装，所以ZooKeeper只有一个节点：

```
zookeeper.connect=hadoop201:2181
```

配置环境变量：

```
$ vim /etc/profile
export KAFKA_HOME=/app/kafka-2.12.0_2.7.0
export PATH=$PATH:$KAFKA_HOME/bin
```

让环境变量生效：

```
$source /etc/profile
```

查看Kafka的版本：

```
[hadoop@server201 bin]$ kafka-server-start.sh --version
[2021-03-26 18:56:35,340] INFO Registered kafka:type=kafka.Log4jController MBean (kafka.utils.Log4jControllerRegistration$)
2.7.0 (Commit:448719dc99a19793)
```

步骤03 启动Kafka。

使用 kafka-server-start.sh启动Kafka进程：

```
$ kafka-server-start.sh /app/kafka-2.0/config/server.properties &
```

建议使用--daemon参数启动：

```
$ bin/kafka-server-start.sh -daemon config/server.properties
```

查看进程：

```
[hadoop@server201 app]$ jps
1381 QuorumPeerMain
1894 Kafka
```

查看ZooKeeper里面的节点，可以看到cluster、brokers已经注册到ZooKeeper里面了：

```
[zk: localhost:2181(CONNECTED) 0] ls /
[cluster, controller_epoch, controller, brokers, zookeeper, admin, isr_change_notification, consumers, log_dir_event_notification, latest_producer_id_block, config]
[zk: localhost:2181(CONNECTED) 1] ls /brokers
[ids, topics, seqid]
[zk: localhost:2181(CONNECTED) 2] ls /brokers/topics
[]
```

步骤04 注册一个Topic。

kafka-topics.sh可以进行创建、删除Topic等操作。通过--help可以查看此命令的帮助：

```
[hadoop@server201 bin]$ kafka-topics.sh --help
This tool helps to create, deslete, describe, or change a topic.
Option                                   Description
------                                   -----------
--alter                                  Alter the number of partitions,
                                         replica assignment, and/or
...
```

注册或创建一个Topic：

```
$ kafka-topics.sh --create --bootstrap-server server201:9092--topic mytopic \
> --replication-factor 1 --partitions 1
```

```
Created topic "mytopic".
```

查看这个Topic：

```
$ kafka-topics.sh --bootstrap-server server201:9092 --describe
Topic:mytopic    PartitionCount:1    ReplicationFactor:1    Configs:
Topic: mytopic    Partition: 0    Leader: 0    Replicas: 0    Isr: 0
```

或使用list列出所有Topic：

```
[hadoop@server201 bin]$ kafka-topics.sh --zookeeper server201:2181 --list
mytopic
```

Topic在Kafka中是主题的意思，生产者将消息发送到主题，消费者再订阅相关的主题，并从主题上拉取消息。在创建Topic的时候，有两个参数是需要填写的，那就是Partitions和replication-factor。

1. Partitions

主题分区数。在创建Topic时，通过--partitions来指定。

Kafka通过分区策略，将不同的分区分配在一个集群中的Broker上，一般会分散在不同的Broker上。当只有一个Broker时，所有的分区就只分配到该Broker上。

消息会通过负载均衡发布到不同的分区上，消费者会监测偏移量来获取哪个分区有新数据，从而从该分区上拉取消息数据。

分区数越多，在一定程度上会提升消息处理的吞吐量，因为Kafka是基于文件进行读写，因此也需要打开更多的文件句柄，从而增加一定的性能开销。

如果分区过多，那么日志分段也会很多，写的时候由于是批量写，其实就会变成随机写了，随机I/O这个时候对性能影响很大。所以一般来说Kafka不能有太多的Partition。

图9-2设置topic-1的Partitions为2，会自动分配在不同的Broker上，采用均匀分配策略，当Broker和Partition个数一样时，就均匀分布在不同的Broker上。这个图示描述了当Partition与Broker个数相同时的分配情况。

图9-3描述了Partition与Broker个数不相同的分配情况（建议不要出现这种情况），topic-1的Partitions个数为3，但Broker的个数为2。

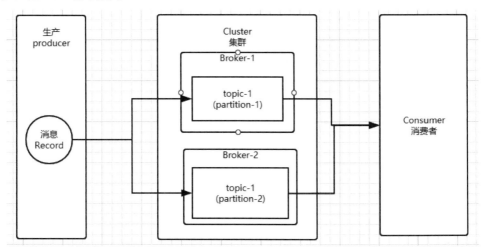

图9-2　设置Partition与Broker个数相同时的分配情况

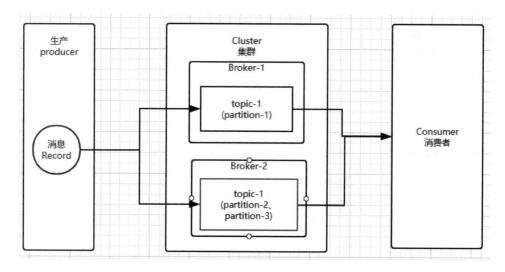

图9-3 设置Partition与Broker个数不相同时的分配情况

2. replication-factor

用来设置主题的副本数。每个主题可以有多个副本，副本位于集群中不同的Broker上，也就是说，副本的数量不能超过Broker的数量，否则创建主题时会失败。

假如我们目前已经有三个Broker（即集群节点数量为3），则创建以下Topic：

```
$kafka-topics.sh --create --topic topicA \
--boostrap-server server202,server203,server201:9092 \
 --replication-factor 2 --partitions 3
Created topic topicA.
```

上面的脚本创建了一个topicA，指定这个Topic的每一个分区副本数量为2，则查看这个topicA的信息为：

```
$ kafka-topics.sh --describe \
--boostrap-server server202,server203,server201:9092 --topic topicA \
topicA PartitionCounter:3    ReplicationFactor:2 Configs:
 Topic:topicA  Partition:0       Leader:201    Replicas:201,203    Isr:201,203
 Topic:topicA  Partition:0       Leader:202    Replicas:201,202    Isr:201,202
 Topic:topicA  Partition:0       Leader:203    Replicas:202,203    Isr:203,202
```

可见，每一个分区（如partition:0）在集群的不同broker上保存了两个副本。

再来测试一个：

如果在集群为3的Broker上指定partitions=1和replication-factor=1，则只会在一个Broker上创建Partition且只会有一个副本，这样做很不合理。

创建：

```
[hadoop@server101 ~]$ kafka-topics.sh --create --topic topicB --boostrap-server server101,server102,server103:9092 --replication-factor 1 --partitions 1
    Created topic topicB.
```

查看结果可见，虽然在集群上，但却保存到一个Borker中：

```
[hadoop@server101 ~]$ kafka-topics.sh --describe \
```

```
--boostrap-server server101,server102,server103:9092 --topic topicB
    Topic:topicB    PartitionCount:1    ReplicationFactor:1    Configs:
    Topic: topicB   Partition: 0    Leader: 101    Replicas: 101    Isr: 101
```

最后的建议：

- replication-factor应该小于或等于Broker的数量。
- Partitions等于Broker的数量。

步骤05 创建发布者和消费者。

创建发布者，并发布一些数据：

```
[hadoop@server201 app]$ kafka-console-producer.sh \
--broker-list server201:9092 --topic mytopic
    >Jack Mary
    >rose
    >Jack
```

创建消费者，将会读取发布者发布的数据，并显示到控制台：

```
[hadoop@server201 app]$ kafka-console-consumer.sh \
--bootstrap-server server201:9092 \
    > --topic mytopic --from-beginning
    Jack Mary
    rose
    Jack
```

步骤06 使用Java代码访问topic。

请确定之前名称为mytopic的Topic依然存在，可以通过--list查看。

```
[hadoop@server201 bin]$ kafka-topics.sh --boostrap-server server201:2181 --list
mytopic
```

添加Kafka的依赖：

```xml
<dependency>
    <groupId>org.apache.kafka</groupId>
    <artifactId>kafka_2.12</artifactId>
    <version>2.7.0</version>
</dependency>
```

创建发布者，启动，并输入一些数据，完整代码如代码9.1所示。

代码9.1 MyProducer.java

```
01 package org.hadoop.kakfa;
02 import org.apache.kafka.clients.producer.KafkaProducer;
03 import org.apache.kafka.clients.producer.Producer;
04 import org.apache.kafka.clients.producer.ProducerConfig;
05 import org.apache.kafka.clients.producer.ProducerRecord;
06 import org.apache.kafka.common.serialization.StringSerializer;
07 import java.util.Properties;
08 import java.util.Scanner;
09 public class MyProducer {
10     public static void main(String[] args) throws Exception {
11         Properties properties = new Properties();
```

```
12        properties.put("bootstrap.servers", "server201:9092");
13        //以下是可选属性
14        properties.put("acks", "all");
15        properties.put("retries", 0);
16        properties.put("batch.size", 16384);
17        properties.put("linger.ms", 1);
18        properties.put("buffer.memory", 33554432);
19        //key.serializer=org.apache.kafka.common.serialization.StringSerializer
20        properties.put(ProducerConfig.KEY_SERIALIZER_CLASS_CONFIG,
    StringSerializer.class.getName());
21        //value.serializer=org.apache.kafka.common.serialization.StringSerializer
22        properties.put(ProducerConfig.VALUE_SERIALIZER_CLASS_CONFIG,
    StringSerializer.class.getName());
23        Producer<String, String> producer = new KafkaProducer<>(properties);
24        Scanner sc = new Scanner(System.in);//接收用户的输入
25        for (int i = 0; ; i++) {
26            String line = sc.nextLine();
27            if (line.equals("exit")) {
28                break;
29            }
30            ProducerRecord<String, String> record =
31                    new ProducerRecord<>("mytopic", "key", line);
32            //发送数据
33            producer.send(record);
34        }
35        producer.close();
36    }
37 }
```

启动后, 我们输入一些数据:

```
This is my First Line
This is My 第二行数据
```

创建消费者, 即会接收到Producer发送的数据, 查看接收到的数据即可。

代码 9.2 MyConsumer.java

```
01 package org.hadoop.kakfa;
02 import org.apache.kafka.clients.consumer.*;
03 import org.apache.kafka.common.serialization.StringDeserializer;
04 import java.time.Duration;
05 import java.util.Arrays;
06 import java.util.Properties;
07 public class MyConsumer {
08    public static void main(String[] args) {
09        Properties properties = new Properties();
10        properties.put("bootstrap.servers", "server201:9092");
11        //以下是可选属性
12        properties.put("group.id", "test");
13        properties.put("enable.auto.commit", "true");
14        properties.put("auto.commit.interval.ms", "1000");
15        properties.put("session.timeout.ms", "30000");
16        //key.serializer=org.apache.kafka.common.serialization.StringSerializer
17        //配置key, value如何解析出来
```

```
18          properties.put(ConsumerConfig.KEY_DESERIALIZER_CLASS_CONFIG,
   StringDeserializer.class.getName());
19          properties.put("value.deserializer",
   "org.apache.kafka.common.serialization.StringDeserializer");
20          Consumer<String, String> consumer = new KafkaConsumer<String,
   String>(properties);
21          //监听的Topic名称
22          consumer.subscribe(Arrays.asList("mytopic"));
23          while (true) {
24              ConsumerRecords<String, String> records =
   consumer.poll(Duration.ofSeconds(2));
25              for (ConsumerRecord<String, String> record : records) {
26                  System.err.println(">>>>>>>>>>:" + record.key() + "\t" +
   record.value());
27              }
28          }
29      }
30  }
```

运行代码后，可以看到接收到数据如下：

```
>>>>>>>>>>:key    This is my First Line
>>>>>>>>>>:key    This is My 第二行数据
```

9.3.2 集群部署

集群部署的核心是配置config/server.properties文件：

```
broker.id=101  #唯一的标识int类型
listeners=PLAINTEXT://server101:9092 #指定本机的地址
log.dirs=/home/isoft/logs/kafka   #指定一个空的已经存在的目录
#指定外部ZooKeeper的地址，可选的指定一个子目录只保存kafka的信息
zookeeper.connection=server101,server102,server103:2181/kafka
```

配置规划如表9-1所示。

表9-1 集群部署的配置规划

主机/ip	软件	进程	标识
server101 192.168.56.101	ZooKeeper-3.5.5 Kafka_2.11_2.3.1 JDK-1.8	QuorumPeerMan Kafka	myid=101 broker.id=101
server102 192.168.56.102	同上	QuorumPeerMan Kafka	myid=102 broker.id=102
server103 192.168.56.103	同上	QuorumPeerMan Kafka	myid=103 broker.id=103

步骤01 配置ZooKeeper集群并启动。

解压ZooKeeper。

修改ZooKeeper的配置文件$ZOOKEEPER_HOME/conf/zoo.cfg：

```
tickTime=2000
initLimit=10
syncLimit=5
dataDir=/home/isoft/datas/zk
clientPort=2181
admin.serverPort=9999
server.101=server101:2888:3888
server.102=server102:2888:3888
server.103=server103:2888:3888
```

在每台服务器的home/isoft/datas/zk目录下添加myid，内容为当前ZooKeeper的id：

```
$ echo 101 > /home/isoft/datas/zk/myid
```

可以使用脚本，一次启动所有ZooKeeper集群：

```
#!/bin/bash
if [ $# -lt 1 ]; then
    echo "用法: $0 start | stop | status"
    exit 1
fi
hosts=(server101 server102 server103)
cmd=$1
for host in ${hosts[@]};do
    script="ssh ${host} zkServer.sh ${cmd}"
    echo $script
    eval $script
done
exit 0
```

步骤02 安装Kafka。

解压：

```
$ tar -zxvf /home/hadoop/kafka_2.12.0-2.7.0.tgz -C /app/
```

修改名称：

```
$ mv kafka_2.20.0-2.7.0 kafka-2.7.0
```

修改配置文件server.properties：

```
broker.id=101    #根据主机的不同，分别设置broker.id=102,broker.id=103
#根据主机不同，分别设置为server102:9092和server103:9092
listeners=PLAINTEXT://server101:9092
log.dirs=/home/hadoop/logs/kafka
#可以声明一个子目录便于管理
zookeeper.connect=server101,server102,server103:2181/kafka
```

将Kafka分发到其他主机的相同目录下：

```
$ scp -r kafka-2.7.0 server102:/app/
$ scp -r kafka-2.7.0 server103:/app/
```

配置所有主机的环境变量：

```
export KAFKA_HOME=/app/kafka-2.7.0
export PATH=$PATH:$KAFKA_HOME/bin
```

步骤03 启动Kafka。

一台一台地启动Kafka：

```
$ kafka-server-start.sh  /app/kafka-2.7.0/config/server.properties  -daemon
```

使用脚本一次性启动：

```
#!/bin/bash
if [ $# -lt 1 ]; then
    echo "使用方法：$0 start | stop"
    exit 1
fi
cmd=$1
servers=(server101 server102 server103)
if [ $cmd == 'start' ]; then
    echo "启动"
    for host in ${servers[@]};
    do
        script="ssh $host kafka-server-start.sh -daemon /app/kafka-2.7.0/config/server.properties"
        echo $script
        eval $script
    done
    exit 0
elif [ $cmd == 'stop' ]; then
    echo "停止"
    for host in ${servers[@]};do
        script="ssh $host kafka-server-stop.sh /app/kafka-2.7.0/config/server.properties"
        echo $script
        eval $script
    done
    exit 0
else
    echo "错误的参数"
    exit 1
fi
```

Kafka都启动完成以后，查看所有Broker的信息：

版本信息：

```
[hadoop@server101 ~]$ kafka-broker-api-versions.sh --version
2.7.0 (Commit:18a913733fb71c01)
```

输入所有的服务器和输入一个服务器地址返回的信息都是一样的：

```
$ kafka-broker-api-versions.sh \
--bootstrap-server server101:9092,server102:9092,server103:9092
```

以下是输出一个服务器地址：

```
[hadoop@server101 ~]$ kafka-broker-api-versions.sh --bootstrap-server server101:9092
```

返回的信息如下：

```
server103:9092 (id: 103 rack: null) -> (
    Produce(0): 0 to 7 [usable: 7],....
```

```
server101:9092 (id: 101 rack: null) -> (
    Produce(0): 0 to 7 [usable: 7],....
server102:9092 (id: 102 rack: null) -> (
    Produce(0): 0 to 7 [usable: 7],....
```

步骤 04 创建Topic。

创建Topic:

```
$ kafka-topics.sh --create --topic two \
--bootstrap-server server101:9092,server102:9092,server103:9092 \
--partitions 3 --replication-factor 3
```

查看Topic:

```
$ kafka-topics.sh --describe --topic two --bootstrap-server server101:9092
Topic:two       PartitionCount:3        ReplicationFactor:3
Configs:segment.bytes=1073741824
        Topic: two      Partition: 0    Leader: 101     Replicas: 101,103,102   Isr: 101,103,102
        Topic: two      Partition: 1    Leader: 103     Replicas: 103,102,101   Isr: 103,102,101
        Topic: two      Partition: 2    Leader: 102     Replicas: 102,101,103   Isr: 102,101,103
```

步骤 05 发布订阅。

创建发布者:

```
$ kafka-console-producer.sh --topic two \
--broker-list server101:9092,server102:9092,server103:9092
>Jack
>mary
>Rose
```

创建消费者:

```
$ kafka-console-consumer.sh --topic two \
--bootstrap-server server101:9092,server102:9092,server103:9092
--from-beginning
    Rose
    mary
    Jack
```

9.4 小　　结

- Kafka安装。
- Kafka的主要概念：Broker、Topic、Producer、Consumer。
- 使用Java代码访问Kafka。

第 10 章

影评大数据分析项目实战

主要内容：

- ❖ 网络爬虫的基本架构。
- ❖ 自然语言处理的入门级应用。
- ❖ 基于Hadoop的数据清洗。
- ❖ 使用MapReduce进行数据处理。
- ❖ 使用词云工具生成数据可视化的图像。

本章将通过一个综合案例——影评分析实战项目，介绍大数据分析处理的一般流程。通过项目流程的详细实现，掌握数据分析每个流程节点的关键技术及应用，包括编写爬虫程序进行数据采集、采用分词技术进行评论数据分析、采用MapReduce进行数据处理，以及进行数据结果的可视化呈现。为了项目案例调试方便，本章采用Hadoop本地运行模式，项目代码在读者本地计算机即可运行。

10.1 项目介绍

本项目主要包含大数据环境中数据爬取的应用、自然语言处理的应用以及MapReduce的应用实现电影影评分析程序，帮助学者了解大数据体系架构的开发流程，以及利用现有技术解决生活中遇到的问题。本实验大致流程为通过Jsoup爬取电影影评数据，将爬取的数据通过IKAnalyzer自然语言处理工具进行分词，将分词的数据通过MapReduce进行数据清洗，以获得kumo词云图形化展示所要求的数据格式。从而实现从数据爬取、数据分析到数据可视化的整体流程。

10.2 项目需求分析

（1）通过Jsoup爬取电影影评网站的数据。Jsoup是一个开源Java库，用于解析、提取和操作存储在HTML文档中的数据，既然是提取HTML中的数据进行解析，首先要了解需要爬取的网站页面内容，在浏览器输入https://movie.douban.com/浏览豆瓣电影页面，如图10-1所示。

图10-1　豆瓣电影页面

选择想要分析的电影，本案例以电影《惊奇队长》为例，单击进入该电影的简介页面，将页面向下拉，会看到"惊奇队长的短评"一栏，如图10-2所示。

图10-2　短评栏

单击"全部xxx条"跳转到该电影的短评页面，如图10-3所示。

图10-3　短评页面

将短评页面往下拖拉到最后，看到"后页"按钮，单击后查看地址栏，如图10-4所示。

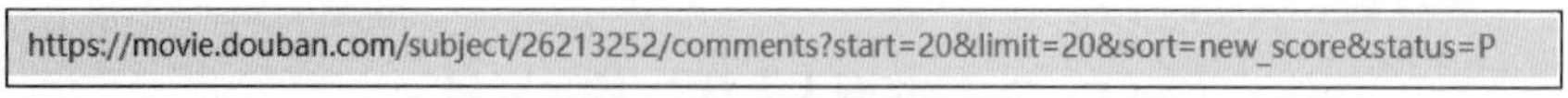

图10-4　后页地址栏

- 26213252：表示电影的id。
- start=20：表示从第几个短评开始。

- limit=20：表示显示多少个短评。

单击"前页"按钮后观察地址栏变化，如图10-5所示。

图10-5 前页地址栏

通过观察地址栏的变化得出结论，改变start的值控制地址栏的变化来获取不同页面的短评数据。在电影短评页面按F12键，进入开发者工具页面查看页面的HTML代码，如图10-6所示。

图10-6 开发者页面

按Ctrl+Shift+C组合键后，开发者工具页面左上角的一个小图标会变成蓝色，如图10-7所示。

图10-7 图标

将鼠标指针拖动到左侧的原页面位置，单击任意一条评论，右侧的开发者工具页面会自动跳转到该条评论对应的HTML代码区域，如图10-8所示。

图10-8 评论源码

通过图10-8可以发现，HTML代码中存储评论数据的是span标签，class为short。单击标签前的小三角展开标签，可查看到HTML中的评论数据，如图10-9所示。

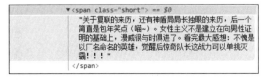

图10-9 HTML中的数据

通过上述分析可得出结论，通过获取每页HTML中class为short的span标签，即可获取每页的评论数据。将这些数据写入到一个文本文件中，如图10-10所示。

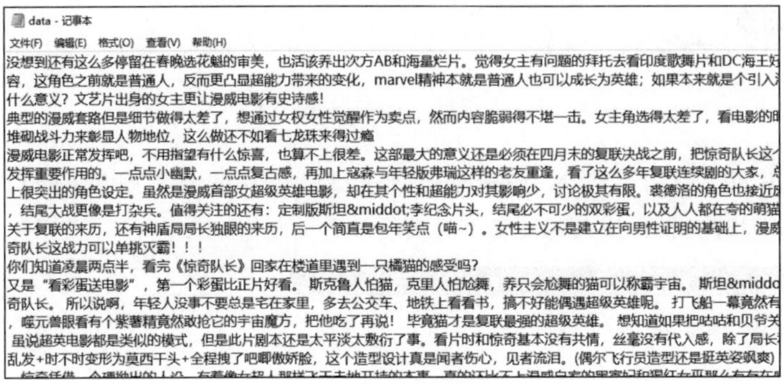

图10-10 评论数据

（2）在上一步爬取完数据的基础上，对影评数据文本利用分词器进行自然语言处理。本案例使用的工具是IKAnalyzer。

IKAnalyzer是一个开源的、基于Java语言开发的、轻量级的中文分词工具包。从2006年12月推出1.0版开始，IKAnalyzer已经推出了4个大版本。最初，它是以开源项目Lucene为应用主体、结合词典分词和文法分析算法的中文分词组件。从3.0版本开始，IK发展为面向Java的公用分词组件，独立于Lucene的项目，同时提供了对Lucene的默认优化实现。在2012版本中，IK实现了简单的分词歧义排除算法，标志着IK分词器从单纯的词典分词向模拟语义分词衍化。

我们将爬取的影评数据进行分词处理，提取特征词语，对分词后的数据进一步分析。例如：

文本原文：

IK-Analyzer是一个开源的，基于java语言开发的轻量级的中文分词工具包。从2006年12月推出1.0版开始，IKAnalyzer已经推出了3个大版本。

分词结果：

ik-analyzer | 是 | 一个 | 一 | 个 | 开源 | 的 | 基于 | java | 语言 | 开发 | 的 | 轻量级 | 量级 | 的 | 中文 | 分词 | 工具包 | 工具 | 从 | 2006 | 年 | 12 | 月 | 推出 | 1.0 | 版 | 开始 | ikanalyzer | 已经 | 推出 | 出了 | 3 | 个大 | 个 | 版本

- 停用词：在分词的结果中可能会出现大量的无用词汇，例如上边的分词结果中出现了"是"这个词，如果在分词结果中不想看到这样的分词，我们可以在分词程序中添加停用词表，那么程序就会在分词时自动过滤这些在停用词表中记录的词汇。本项目我们使用哈工大提供的停用词表。
- 拓展词：如果文章中出现了生僻词汇，那么默认的分词程序是不会将这些词汇单独分出来，我们需要在程序中添加一个拓展词表，例如上边提到的例子，如果我们想将"是一个"划分为一个词，那么在拓展词表中就要加入这个词汇，便于程序去读取。

最终将分词后的数据输出到结果文件中，如图10-11所示。

（3）接下来通过MapReduce程序对分词后的结果进行统计并格式化输出。这里我们需要在MapReduce程序中设置两个job，将默认Key Value输出格式设置为Key:Value的输出格式，其中一个job用来完成词汇的统计，统计分词结果中每个词汇出现的次数，如图10-12所示。

此时的数据是按照MapReduce默认排序输出的结果，后续词云展示用到的数据格式为“1:第一战”，此时的输出结果并不符合我们的需求，所以对第一次job输出的结果文件进行第二次job处理，将第一次输出的Key和Value调换位置，并且将出现次数多的词汇排在最前边，对Key进行逆序排序，为了方便最后词云展示的效果，将数据中单个词汇进行过滤。第二次job的输出结果如图10-13所示。

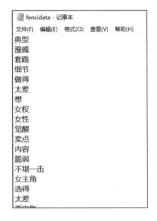

fencidata - 记事本
文件(F) 编辑(E) 格式(O) 查看(V) 帮助(H)
典型
漫威
套路
细节
做得
太差
想
女权
女性
觉醒
卖点
内容
脆弱
不堪一击
女主角
选得
太差

图10-11　分词结果

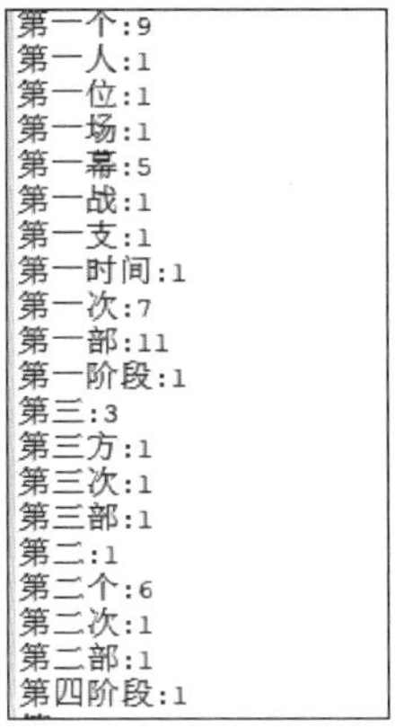
第一个:9
第一人:1
第一位:1
第一场:1
第一幕:5
第一战:1
第一支:1
第一时间:1
第一次:7
第一部:11
第一阶段:1
第三:3
第三方:1
第三次:1
第三部:1
第二:1
第二个:6
第二次:1
第二部:1
第四阶段:1

图10-12　第一次job

195:漫威
140:电影
130:惊奇队长
87:英雄
68:一个
68:女主
66:剧情
64:复联
63:女性
63:角色
59:宇宙
58:彩蛋
54:超级
54:真的

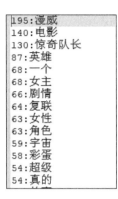

图10-13　第二次job

（4）最终将MapReduce的输出结果通过Kumo程序生成词云，Kumo的目标是利用Java创建功能强大且用户友好的Word Cloud API。Kumo可以直接生成一个图像文件。Word Cloud是关键词的视觉化描述，用于汇总用户生成的标签或一个网站的文字内容。最终输出关于电影《惊奇队长》的词云图像如图10-14所示。

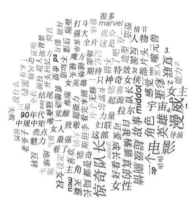

图10-14　词云图

10.3　项目详细实现

10.3.1　搭建项目环境

打开Eclipse，选择 File→New→Other→Maven Project，创建Maven工程，如图10-15所示，单击Next按钮，会进入“New Maven Project”对话框，如图10-16所示。

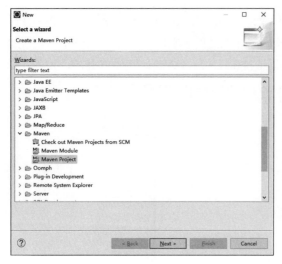 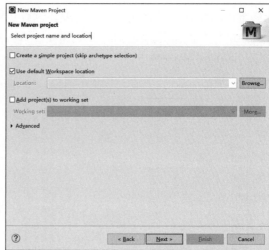

图10-15 创建Maven工程　　　　　图10-16 "New Maven Project"对话框

在图10-16所示的对话框中，勾选"Create a simple project (skip archetype selection)"表示创建一个简单的项目（跳过对原型模板的选择），勾选"User default Workspace location"表示使用本地默认的工作空间（本例我们勾选此项），单击Next按钮，弹出如图10-17所示的工程配置对话框。

图10-17 创建Maven工程配置

在图10-17所示的对话框中，Group Id也就是项目组织唯一的标识符，实际对应Java的包结构，这里输入com.mrchi。Artifact Id就是项目的唯一标识符，实际对应项目的名称，就是项目根目录的名称，这里输入experiment。打包方式选择jar即可，后续创建Web工程选择war包。

单击Finish按钮，此时Maven工程已经创建好了，会发现在Maven项目中有一个pom.xml的配置文件，这个文件就是对项目进行管理的核心配置文件。

本案例需要用到hadoop-common、hadoop-client、jsoup、kumo-core、kumo-tokenizers、ikanalyzer六种依赖，具体代码如下。

代码 10.1　pom.xml

```
01  <project xmlns="http://maven.apache.org/POM/
02  4.0.0" xmlns:xsi="http://www.w3.org/2001/
03  XMLSchema-instance" xsi:schemaLocation="http://maven.apache.org/POM/4.0.0
    http://maven.apache.org/xsd/maven-4.0.0.xsd">
04      <modelVersion>4.0.0</modelVersion>
05      <groupId>com.mrchi</groupId>
06      <artifactId>hadoopInstance</artifactId>
07      <version>0.0.1-SNAPSHOT</version>
08
09      <dependencies>
10          <dependency>
11              <groupId>org.jsoup</groupId>
12              <artifactId>jsoup</artifactId>
13              <version>1.7.3</version>
14          </dependency>
15          <dependency>
16              <groupId>com.kennycason</groupId>
17              <artifactId>kumo-core</artifactId>
18              <version>1.17</version>
19          </dependency>
20          <dependency>
21              <groupId>com.kennycason</groupId>
22              <artifactId>kumo-tokenizers</artifactId>
23              <version>1.17</version>
24          </dependency>
25          <dependency>
26              <groupId>com.github.magese</groupId>
27              <artifactId>ik-analyzer</artifactId>
28              <version>7.7.1</version>
29          </dependency>
30          <dependency>
31              <groupId>org.apache.hadoop</groupId>
32              <artifactId>hadoop-common</artifactId>
33              <version>2.7.4</version>
34          </dependency>
35          <dependency>
36              <groupId>org.apache.hadoop</groupId>
37              <artifactId>hadoop-client</artifactId>
38              <version>2.7.4</version>
39          </dependency>
40      </dependencies>
41  </project>
```

当添加依赖完毕后，Hadoop相关jar包就会自动下载，部分jar包如图10-18所示。

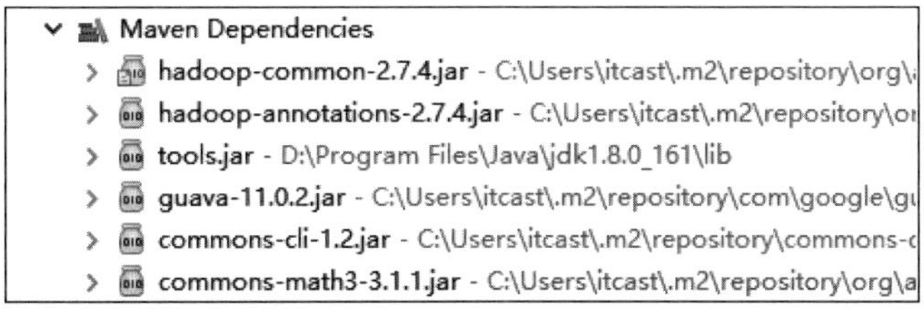

图10-18　成功导入jar包

10.3.2 编写爬虫类

首先在项目src文件夹下创建com.mrchi.move包,在该路径下编写爬虫类moveJsoup,代码实现如下所示。

代码10.2　moveJsoup.java

```java
01 public class moveJsoup {
02     public static void main(String[] args) throws InterruptedException,IOException{
03         int num = 0;
04         //爬取数据存储到本地的地址
05         File fileNmame = new File("D:\\MoveData\\data.txt");
06         BufferedWriter out = new BufferedWriter(new FileWriter(fileNmame));
07         //爬取25页的影评数据,每页20条数据
08         for (int i = 0; i < 25; i++) {
09             String strNum = Integer.toString(num);
10             //爬取数据的url地址
11             String url =
12                 "https://movie.douban.com/subject/26213252/comments?start="+    num
     +"&amp;limit=20&amp;sort=new_score&amp;status=P";
13             Connection connect = Jsoup.connect(url);
14             Document document =
15                 connect.userAgent("浏览器的User-Agent")
16                     .cookie("Cookie", "自己的cookie")
17                         //超时请求的时间
18                     .timeout(6000)
19                         //解析响应时忽略文档的Content-Type
20                     .ignoreContentType(true)
21                         //以GET身份执行请求,并解析结果
22                     .get();
23             //获取span标签中class等于short的内容
24             Elements elements = document.select("span[class=short]");
25             for (Element e : elements) {
26                 //将页面中的内容循环读取写入到文件data.txt中
27                 out.write(e.toString().replaceAll("</?[^>]+>", "")+"\r\n");
28             }
29             num = Integer.parseInt(strNum);
30             num+=20;
31         }
32         //将缓冲区的数据输出
33         out.flush();
34         //关闭连接
35         out.close();
36     }
37 }
```

获取User-Agent和Cookie的方法:使用自己的账号密码登录豆瓣网后,按照案例需求中提到的步骤进入到全部短评页面,按F12键进入到开发者工具模式,然后单击地址栏中的Network,如图10-19所示。

按F5键刷新页面,在Network页面下会出现一个文件,单击该文件会显示详细内容,如图10-20所示。

将文件内容拉到最底部,可以看到Cookie和User-Agent的内容如图10-21所示,将":"号后的内容粘贴到代码中对应的位置即可。

第 10 章　影评大数据分析项目实战 | 207

图10-19　Network页面

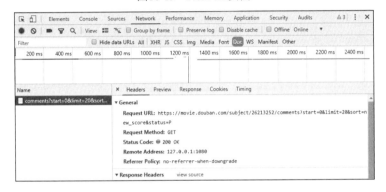

图10-20　文件内容

图10-21　Cookie和User-Agent

注意：Cookie复制到代码中时会报错，因为Cookie中包含""　""双引号，处理时在双引号的前边加上"\"转义符即可，比如：\"asdasd\"。

如果不想使用用户名密码登录豆瓣网，可以将代码中25换成10，因为不登录账号的话只能查看200条评论数据，将代码中.Cookie()一行删除即可。User-Agent在不登录的情况下，也可以用相同的方式获取。第27行的正则是去除字符串中类似于<em></em>的标签，\r\n是对每一条数据间做换行处理。

10.3.3　编写分词类

在com.mrchi.move包下编写分词类moveFenci，代码实现如下所示。

代码10.3　moveFenci.java

```
01 public class moveFenci {
02     public static void main(String[] args) throws IOException {
```

```
03      //创建分词器对象
04      Analyzer analyzer = null;
05      //创建TokenStream流对象,分词的所有信息都会从TokenStream流中获取
06      TokenStream ts = null;
07      //创建InputStreamReader对象,将字节流转换为字符流
08      InputStreamReader reader = null;
09      //创建BufferedWriter对象,将文本写入字符输出流
10      BufferedWriter writer = null;
11      //创建BufferedReader对象,从字符输入流中读取文本
12      BufferedReader br = null;
13
14      try{
15          //分词结果输出的路径
16          File fenCidata = new File("D:\\MoveData\\fencidata.txt");
17          //影评信息输入的路径
18          File moveData = new File("D:\\MoveData\\data.txt");
19          //读取文件的数据,将字节流向字符流的转换
20          reader = new InputStreamReader(new FileInputStream(moveData));
21          //从字符流中读取文本
22          br = new BufferedReader(reader);
23          //创建了一个字符写入流的缓冲区对象,并和指定要被缓冲的流对象相关联
24          writer = new BufferedWriter(new FileWriter(fenCidata));
25          String line = "";
26          //读取一行数据
27          line = br.readLine();
28          //判断文本是否有数据
29          while (line != null) {
30              line = br.readLine();
31              if(line == null){
32                  continue;
33              }
34              //构造分词对象,true表示使用智能分词
35              analyzer = new IKAnalyzer(true);
36              //""指文档的域名,一片文档包含多个域名,这里任意指定即可
37              ts = analyzer.tokenStream("", line);
38              //获取每个单词信息
39              CharTermAttribute term=ts.getAttribute(CharTermAttribute.class);
40              //TokenStream的流程需包含reset()重置
41              ts.reset();
42              //遍历分词数据
43              while(ts.incrementToken()){
44                  //将数据写入到缓冲区中
45                  writer.write(term.toString()+"\r\n");
46              }
47
48          }
49      }
50      catch(Exception e){
51          e.printStackTrace();
52      }finally{
53          //关闭流
54          analyzer.close();
55          //关闭流
56          reader.close();
57          //关闭流
58          br.close();
```

```
59                //刷新该流中的缓冲,将缓冲数据写到目标文件中去
60                writer.flush();
61                //关闭此流,在关闭前会先刷新它
62                writer.close();
63           }
64      }
65 }
```

使用IKAnalyzer内部的智能分词方法对文本文件进行分词处理,将分词后的结果输出到文本文件中,每个词汇以\r\n换行符进行分隔。在本案例的分词过程中,使用到了拓展词库和停用词库。具体配置方法如下:

查看配置文件的目录结构,如图10-22所示。

首先在src/main/resources目录下新建XML文件,选中src/main/resources目录右键File→New→Other,如图10-23所示,选择XML File,单击Next按钮。

图10-22　目录结构

将XML文件命名为IKAnalyzer.cfg.xml(在File name后的输入框内输入),然后单击Finish按钮完成创建,创建完成后打开该XML文件,选中IKAnalyzer.cfg.xml文件右击依次打开Open With→Text Editor,向XML文件中写入如下内容:

```xml
<?xml version="1.0" encoding="UTF-8"?>
<!DOCTYPE properties SYSTEM "http://java.sun.com/dtd/properties.dtd">
<properties>
    <comment>IK Analyzer 扩展配置</comment>
    <!--用户可以在这里配置自己的扩展字典 -->
    <entry key="ext_dict">ext.dic;</entry>
    <!--用户可以在这里配置自己的扩展停止词字典-->
    <entry key="ext_stopwords">stopword.dic;</entry>
</properties>
```

XML文件中指定扩展词和停用词文本的名称。

在src/main/resources目录下新建拓展词File文件,选中src/main/resources目录右击依次选择File→New→Other,如图10-24所示,选择File,单击Next按钮。

图10-23　选择 XML File

图10-24　选择 File

将File文件命名为ext.dic（在File name后的输入框内输入），然后单击Finish按钮完成创建，创建完成后打开该File文件，选中ext.dic文件右击依次选择Open With→Text Editor，向File文件中写入想要添加的拓展词汇，如图10-25所示。

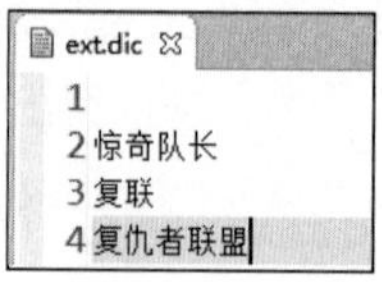

通常情况下，分词器默认会把复和联分开处理，如果我们想让这两个字组合显示，则需要向分词器提供这两个字组合的词汇。

图10-25　拓展词

在src/main/resources目录下新建停用词File文件，命名为stopword.dic，向文件中写入想要添加的停用词汇，如图10-26所示。

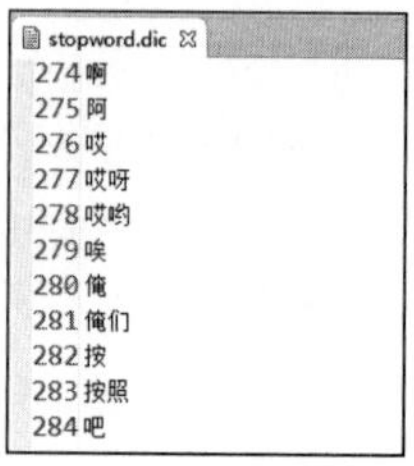

分词器在分词的过程中会自动去除停用词表中提到的词汇。本案例用到的是哈工大的停用词表，读者也可以根据实际情况自行设计。

图10-26　停用词

10.3.4　第一个job的Map阶段实现

在com.mrchi.move包下新建Mapper类moveMapper，代码实现如下所示。

代码 10.4　moveMapper.java

```java
01 public class moveMapper extends Mapper<LongWritable, Text, Text, IntWritable> {
02     private final IntWritable count= new IntWritable(1);
03     private Text inkey = new Text();
04     @Override
05     protected void map(LongWritable key, Text value, Context context)
06         throws IOException, InterruptedException {
07         String word = value.toString();
08         inkey.set(word);
09         context.write(inkey,count);
10     }
11 }
```

Map阶段获取分词结果文件中的每一个词汇，将这些词汇以"Key,Value"的形式进行输出，Key为每个词汇，Value为1。

10.3.5　第一个job的Reducer阶段实现

在com.mrchi.move包下新建Reducer类moveReducer，代码实现如下所示。

代码 10.5　moveReducer.java

```java
01 public class moveReducer extends Reducer<Text, IntWritable,Text,IntWritable> {
02 
03     @Override
04     protected void reduce(Text key, Iterable<IntWritable> values,Context context)
       throws IOException, InterruptedException {
05         int total = 0;
06         for (IntWritable val : values) {
07             total = total + val.get();
08         }
09         context.write( key,new IntWritable(total));
10     }
11 }
```

将shuffle阶段传过来的数据进行处理,shuffle会将相同Key的Value放在一起,Reduce将Value遍历后聚合,从而得出词汇(Key)在所有评论中出现的次数(Value)。

10.3.6 第二个job的Map阶段实现

在com.mrchi.move包下新建Mapper类moveSortMapper,代码实现如下所示。

代码10.6　moveSortMapper.java

```java
01 public class moveSortMapper extends Mapper<LongWritable, Text, IntWritable, Text> {
02     private final static IntWritable wordCount = new IntWritable();
03     private Text word = new Text();
04     @Override
05     protected void map(LongWritable key, Text value, Context context)
06             throws IOException, InterruptedException {
07         //以":"作为分隔符处理每一行数据
08         StringTokenizer tokenizer =
09                     new StringTokenizer(value.toString(),":");
10         while (tokenizer.hasMoreTokens()) {
11             //将分隔后的第一个值赋值给变量a
12             String a = tokenizer.nextToken().trim();
13             word.set(a);
14             //将分隔后的第二个值赋值给变量b
15             String b = tokenizer.nextToken().trim();
16             //将b转成Integer类型
17             wordCount.set(Integer.valueOf(b));
18             context.write(wordCount, word);
19         }
20     }
21 }
```

读取第一个job的输出,处理每一行数据,使map输出为"Key,Value"形式,Key(词汇出现的次数)为IntWritable类型,Value(词汇)为Text类型。

10.3.7 第二个job的自定义排序类阶段的实现

在com.mrchi.move包下编写自定义排序类moveSortComparator继承WritableComparator,代码实现如下所示。

代码10.7　moveSortComparator.java

```java
01 public class moveSortComparator extends WritableComparator{
02     protected moveSortComparator() {
03         super(IntWritable.class, true);
04     }
05     @Override
06     public int compare(WritableComparable a, WritableComparable b) {
07         return -super.compare(a, b);
08     }
09 }
```

Reducer默认排序是从小到大（数字），而我们期望出现次数多的词语排在前面，所以需要重写排序类WritableComparator。

10.3.8 第二个 job 的自定义分区阶段实现

在com.mrchi.move包下编写自定义分区类moveSortPartition继承Partitioner，代码实现如下所示。

代码 10.8   moveSortPartition.java

```
01  import org.apache.hadoop.mapreduce.Partitioner;
02
03  public class moveSortPartition <K, V> extends Partitioner<K, V> {
04      @Override
05      public int getPartition(K key, V value, int numReduceTasks) {
06          int maxValue = 50;
07          int keySection = 0;
08          // 只有传过来的key值大于maxValue，并且numReduceTasks大于1个才需要分区，否则直接返回0
09          if (numReduceTasks > 1 && key.hashCode() < maxValue) {
10              int sectionValue = maxValue / (numReduceTasks - 1);
11              int count = 0;
12              while ((key.hashCode() - sectionValue * count) > sectionValue) {
13                  count++;
14              }
15              keySection = numReduceTasks - 1 - count;
16          }
17          return keySection;
18      }
19  }
```

如果有多个Reducer任务，Reducer的默认排序只是对发送到该Reducer下的数据局部排序。如果想达到全局排序，需要我们手动去写Partitioner。Partitioner的作用是根据不同的Key，制定相应的规则分发到不同的Reducer中。

10.3.9 第二个 job 的 Reduce 阶段实现

在com.mrchi.move包下编写Reducer类moveSortReducer，代码实现如下所示。

代码 10.9   moveSortReducer.java

```
01  public class moveSortReducer extends Reducer<IntWritable, Text, IntWritable, Text> {
02      private Text result = new Text();
03
04      @Override
05      protected void reduce(IntWritable key, Iterable<Text> values,Context context)
        throws IOException, InterruptedException {
06          for (Text val : values) {
07              if(val.toString().length() >1){
08                  result.set(val.toString());
09                  context.write(key, result);
10              }
11          }
12      }
13  }
```

将shuffle阶段传过来的数据进行处理，遍历相同Key的Value，第7行去除Value为一个词的值，为了使词云结果展示效果更佳。

10.3.10 Run 程序主类实现

在com.mrchi.move包下编写Main类moveRun，代码实现如下所示。

代码 10.10　moveRun.java

```
01 public class moveRun {
02     public static void main(String[] args) throws IllegalArgumentException,
   IOException, ClassNotFoundException, InterruptedException {
03         BasicConfigurator.configure();
04         Configuration conf = new Configuration();
05         //定义mapreduce输出key,value的分隔符
06         conf.set("mapred.textoutputformat.separator", ":");
07         Job job1 = new Job(conf, "count");
08         job1.setJarByClass(moveRun.class);
09         //第一次job的mapper
10         job1.setMapperClass(moveMapper.class);
11         //第一次job的reduce
12         job1.setReducerClass(moveReducer.class);
13         //第一次job输出key的文件类型
14         job1.setOutputKeyClass(Text.class);
15         //第一次job输出value的文件类型
16         job1.setOutputValueClass(IntWritable.class);
17         //第一次job输入文件的地址
18         FileInputFormat.addInputPath(job1, new
   Path("D:\\MoveData\\fencidata.txt"));
19         //第一次job输出结果的地址
20         FileOutputFormat.setOutputPath(job1, new Path("D:\\MoveData\\result1"));
21         job1.waitForCompletion(true);
22 
23         Job job2 = new Job(conf, "sort");
24         job2.setJarByClass(moveRun.class);
25         //map输出key的类型
26         job2.setMapOutputKeyClass(IntWritable.class);
27         //map输出value的类型
28         job2.setMapOutputValueClass(Text.class);
29         //输出key的类型
30         job2.setOutputKeyClass(IntWritable.class);
31         //输出value的类型
32         job2.setOutputValueClass(Text.class);
33         //定义mapper类
34         job2.setMapperClass(moveSortMapper.class);
35         //定义reducer类
36         job2.setReducerClass(moveSortReducer.class);
37         //自定义排序类
38         job2.setSortComparatorClass(moveSortComparator.class);
39         //自定义分区类
40         job2.setPartitionerClass(moveSortPartition.class);
41         //输入文件的地址
42         FileInputFormat.addInputPath(job2, new Path("D:\\MoveData\\result1"));
43         //输出文件的地址
44         FileOutputFormat.setOutputPath(job2,
```

```
45  new Path("D:\\MoveData\\result2"));
46          System.exit(job2.waitForCompletion(true) ? 0 : 1);
47      }
48  }
```

设置MapReduce工作任务的相关参数，本案例采用本地运行模式，对指定的本地D:\\MoveData\\fencidata.txt目录下的分词结果和第一次job的输出D:\\MoveData\\result1实现数据清洗，并将结果最终输入到本地D:\\MoveData\\result2目录下，设置完毕后，运行程序即可。

10.3.11　编写词云类

在com.mrchi.move包下编写词云类moveShow，代码实现如下所示。

代码 10.11　moveShow.java

```
01  public abstract class moveShow {
02      public static void main(String[] args) throws IOException {
03          BasicConfigurator.configure();
04          //建立词频分析器，设置词频，以及词语最短长度，此处的参数配置视情况而定即可
05          FrequencyAnalyzer frequencyAnalyzer = new FrequencyAnalyzer();
06          //选取600个词
07          frequencyAnalyzer.setWordFrequenciesToReturn(600);
08          //引入中文解析器
09          frequencyAnalyzer.setWordTokenizer(new ChineseWordTokenizer());
10          //指定文本文件路径，生成词频集合
11          FrequencyFileLoader ffl = new FrequencyFileLoader();
12          List<WordFrequency> wordFrequencyList=frequencyAnalyzer
13  .loadWordFrequencies(ffl.load(new File("D:\\MoveData\\result2\\part-r-00000")));
14
15          //设置图片分辨率
16          Dimension dimension = new Dimension(1920,1080);
17          //此处的设置采用内置常量即可，生成词云对象
18          WordCloud wordCloud =
19                  new WordCloud(dimension,CollisionMode.PIXEL_PERFECT);
20          //设置边界及字体
21          wordCloud.setPadding(2);
22          java.awt.Font font = new java.awt.Font("STSong-Light", 2, 20);
23          //设置词云显示的三种颜色，越靠前设置表示词频越高的词语的颜色
24          wordCloud.setColorPalette(new LinearGradientColorPalette(Color.RED,
    Color.BLUE, Color.GREEN, 30, 30));
25          wordCloud.setKumoFont(new KumoFont(font));
26          //设置背景色
27          wordCloud.setBackgroundColor(new Color(255,255,255));
28          //设置背景图层为圆形
29          wordCloud.setBackground(new CircleBackground(255));
30          wordCloud.setFontScalar(new SqrtFontScalar(12, 45));
31          //生成词云
32          wordCloud.build(wordFrequencyList);
33          wordCloud.writeToFile("D:\\MoveData\\move.png");
34
35      }
36
37  }
```

读取MapReduce最终输出的结果，选取结果文件中前600个词汇组成词云，因为MapReduce的第二次job做过从大到小的排序，所以选取的会是出现次数较多的前600个，根据Key（出现次数）大小在词云中通过词汇字体的大小来体现。

10.3.12 效果测试

为了保证程序正常执行，需要在本地创建D:\\MoveData\\目录。

首先执行爬虫类，在D:\\MoveData\\目录下会生成data的文件。
接着执行分词类，在D:\\MoveData\\目录下会生成fencidata的文件。
然后执行moveRun类，在D:\\MoveData\\目录下会生成result1和result2的两个文件夹。
最后执行moveShow类，在D:\\MoveData\\目录下会生成move.png图片。
最终D:\\MoveData\\目录下的内容如图10-27所示。

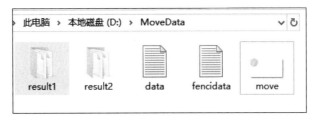

图10-27　效果测试

至此，影评分析案例全部完成，可以通过move文件查看《惊奇队长》这部电影的具体内容。

第 11 章

旅游酒店评价大数据分析项目实战

主要内容：

- Hadoop HDFS数据存储。
- 构建Hive数据仓库表。
- 基于Hadoop MapReduce的数据清洗。
- Hive数据导出到MySQL。
- 基于ECharts可视化展示。

本章将通过一个基于城市旅游酒店的基本数据及用户评论数据的集合，综合运用Hadoop进行大数据分析项目实战。本项目基于山东省青岛市酒店基本数据和用户评论数据，经过HDFS应用程序进行上传和存储；然后运用MapReduce进行数据预处理，利用处理之后的数据构建Hive数据仓库，并结合用户需求生成数据仓库分析结果表，将表导出到MySQL；最后构建基于Spring Boot框架的Web项目，并结合ECharts技术实现数据可视化呈现，为企业和用户提供决策支持。

11.1 项目介绍

随着计算机网络发展，各大型网站及平台信息实时更新，产生了大量数据。在当今大数据背景下，各行各业积累了海量数据，这些数据具有数据容量大、类型多、增长速度快、价值密度高的特点。许多学者也展开了关于大数据分析算法、分析模式及分析软件工具方面的研究。其中，在大数据结构模型和数据科学理论体系、大数据分析和挖掘基础理论方面有很大的进步，大数据的应用领域也从科学、工程、电信等领域扩展到各行各业。许多规模较大的酒店都有自己的酒店管理系统，提供了完善的酒店管理和酒店预订、酒店评价等服务。部分中小型酒店由于缺乏投资，依托第三方平台提供在线服务，客户进行操作后，第三方平台会生成记录保存。酒店长期积累了大量的在线基本数据和用户评论数据。针对酒店行业，如何利用大数据技术来对现有的数据进行处理和分析，帮助酒店从业者和出行用户提供直观的参考决策，这是我们急需解决的问题。一方面，根据用户在线评论数据，帮助酒店从业者提供直观的决策支持，改善酒店管理，以获取最大利润；另一方面，提供某地区的酒店基本满意度情况、酒店分布情况、热门酒店等可视化图表，为用户出行提供可靠的参考。

为了使用户对旅游目标城市的酒店住宿和用户满意度、城市各地区酒店分布、用户出游目的等情况有更加直观、明确的了解，并为用户提前规划好住宿和旅游景点的选择提供决策支持，本项目基于山东省青岛市酒店基本数据和用户评论数据来构建大数据平台及数据仓库，并进行统计分析，最后以Web网页的形式将分析结果和决策以可视化图表方式进行展现。本项目也能为酒店从业者提供一定的决策支持，方便其在前期市场调查过程中提前了解各区酒店分布、满意度、用户出游目的等信息，比如，可以根据用户出游类型占比等信息来为酒店从业者规划酒店类型及相关配套等。

该项目的具体过程为：对青岛市的酒店评论和酒店基本数据进行大数据分析和处理，数据存储到Hadoop集群，经过数据清洗后构建Hive数据仓库，并基于Hive数据仓库进行数据分析，将分析结果最终导入到MySQL中，最后构建基于Java Web的项目进行酒店数据可视化展示。

本项目运行前需要搭建Hadoop基础环境，这里为了项目展开方便，采用了伪分布式的Hadoop集群，服务器的IP地址为192.168.1.110，安装Hadoop、Hive等相关框架或组件，操作系统采用CentOS，MySQL相关软件安装在该服务器上。开发环境在Windows本地主机，使用IDEA作为开发工具。

11.2 项目需求分析

11.2.1 数据集需求

为了给游客提供城市酒店满意度、酒店分布等出行需求，项目对数据集有一定要求。本项目已提供山东省青岛市的酒店基本数据和酒店评论数据两个数据集文件。下面对两个文件的属性做一下说明。

（1）酒店用户评论数据。关于酒店用户评论数据采集的数据格式，描述为：酒店评论的主键（order_id），以便于验证信息的唯一性；酒店网页地址（url），便于查看数据是否采集正确。酒店名字（hotel_name），便于统计酒店数量；酒店评论的发布日期（post_time），便于了解是否是最近发布的；酒店评论的用户名（user_name），便于对评论进行用户画像；本条酒店评论的评论内容（content），这是后面进行酒店顾客意见挖掘的文本数据；酒店评论的用户打分（user_score），这个数据属性能和用户评论进行映射，便于后面的情感分析训练集的制作。

（2）酒店基本信息数据。关于酒店基本数据采集的数据格式，描述为：酒店id、酒店名称、酒店评分、评论人数、用户推荐指数（%）、酒店地址、酒店星级、星级详情。

11.2.2 功能需求

对山东省某个城市进行大数据分析和处理，这里以青岛市酒店数据为例。首先需要把青岛市的酒店基本信息数据和用户评论数据进行预处理（数据清洗）之后，上传到Hadoop平台的HDFS存储，然后基于MapReduce进行二次数据清洗，然后基于HDFS中的两个数据集构建Hive数据仓库，然后基于Hive数据仓库，根据用户关心的酒店及评论信息的维度进行数据分析处理，根据系统功能要求分为5个关注角度，并为每一个关注角度创建Hive内部表，具体如下：

（1）用户印象统计，也是用户对该城市总体满意度的情况。用户对住过酒店发表评论同时也可以打分，我们可以根据用户对该城市的酒店总体评分情况来统计用户总体印象。评分4.5~5分为优良，3.5~4.5为良好，3.5以下为差，统计酒店用户评分等级比例。

（2）统计在线评论数最多的十大酒店、十大网络人气酒店。一般情况下，一家酒店的评论数量能代表这家酒店的人气，这里统计的是酒店名称和评论数目。

（3）不同旅游类型占比统计。根据用户评论Hive外部表hotel_data，进行不同旅游类型的统计，根据旅游类型结合用户满意度情况，为用户旅游出行提供参考，了解该城市更适合哪种类型的旅游。

（4）酒店星级分布情况统计。设计酒店星级和数量两个属性，显示不同星级的酒店数量占比，为不同层次的用户提供星级酒店的数量分布。

（5）城市不同地区的酒店数量分布情况。以热力图方式呈现，同时需要显示每个城市的酒店数量和平均评论得分情况。

最后，将产生的Hive内部表数据，利用Sqoop导出到MySQL数据库。

数据可视化部分，根据用户5个关注角度的分析结果，构建数据展现的Web项目，采用技术是：Spring Boot+MyBatis+MySQL，开发工具使用IDEA，图表采用ECharts来进行页面图表渲染支持。为了提高页面加载速度和用户体验，可采用Ajax异步加载的方式来进行图表呈现。

11.3　项目详细实现

项目的主要开发流程如图11-1所示。

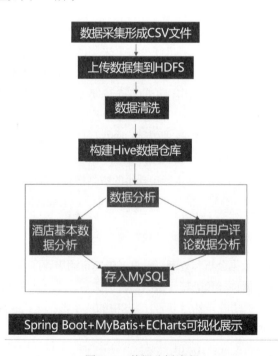

图11-1　数据分析流程

11.3.1 数据集上传到 HDFS

将经过数据预处理后的酒店数据集，以及用户评论数据集对应的CSV文件上传到HDFS存储。这里为了调试代码方便，将程序部分放到Windows系统，Hadoop安装到CentOS系统，具体框架安装部分可以参照之前相关章节的内容。

（1）数据上传程序在Windows系统下，在IDEA下新建一个Maven项目，用于数据上传和数据清洗。具体项目pom.xml代码如下：

代码 11.1　pom.xml

```xml
01 <!--设置依赖版本号-->
02 <properties>
03     <scala.version>2.11.8</scala.version>
04     <hadoop.version>3.2.2</hadoop.version>
05     <spark.version>3.1.1</spark.version>
06 </properties>
07 <dependencies>
08     <!--Scala-->
09     <dependency>
10         <groupId>org.scala-lang</groupId>
11         <artifactId>scala-library</artifactId>
12         <version>${scala.version}</version>
13     </dependency>
14     <!--Hadoop-->
15     <dependency>
16         <groupId>org.apache.hadoop</groupId>
17         <artifactId>hadoop-client</artifactId>
18         <version>${hadoop.version}</version>
19     </dependency>
20     <dependency>
21         <groupId>org.apache.hadoop</groupId>
22         <artifactId>hadoop-common</artifactId>
23         <version>${hadoop.version}</version>
24     </dependency>
25     <dependency>
26         <groupId>org.apache.hadoop</groupId>
27         <artifactId>hadoop-hdfs</artifactId>
28         <version>${hadoop.version}</version>
29     </dependency>
30     <dependency>
31         <groupId>org.apache.hadoop</groupId>
32         <artifactId>hadoop-mapreduce-client-core</artifactId>
33         <version>${hadoop.version}</version>
34     </dependency>
35     <dependency>
36         <groupId>junit</groupId>
37         <artifactId>junit</artifactId>
38         <version>4.12</version>
39     </dependency>
40     <dependency>
41         <groupId>org.apache.zookeeper</groupId>
42         <artifactId>zookeeper</artifactId>
```

```
43          <version>3.4.10</version>
44      </dependency>
45      <dependency>
46          <groupId>com.databricks</groupId>
47          <artifactId>spark-csv_2.10</artifactId>
48          <version>1.0.3</version>
49      </dependency>
50      <!-- https://mvnrepository.com/artifact/com.google.code.gson/gson
51      <dependency>
52          <groupId>com.google.code.gson</groupId>
53          <artifactId>gson</artifactId>
54          <version>2.8.0</version>
55      </dependency>
56       <dependency>
57          <groupId>org.apache.kafka</groupId>
58          <artifactId>kafka_2.11</artifactId>
59          <version>1.0.0</version>
60      </dependency>-->
61
62  </dependencies>
```

(2)编写具体上传文件的Java程序，代码如下：

代码 11.2　HDFS_CRUD.java

```
01  import java.io.FileNotFoundException;
02  import java.io.IOException;
03  import org.apache.hadoop.conf.Configuration;
04  import org.apache.hadoop.fs.BlockLocation;
05  import org.apache.hadoop.fs.FileStatus;
06  import org.apache.hadoop.fs.FileSystem;
07  import org.apache.hadoop.fs.LocatedFileStatus;
08  import org.apache.hadoop.fs.Path;
09  import org.apache.hadoop.fs.RemoteIterator;
10  import org.junit.Before;
11  import org.junit.Test;
12
13  public class HDFS_CRUD {
14      FileSystem fs = null;
15
16      @Before
17      public void init() throws Exception {
18          // 构造一个配置参数对象，设置一个参数：我们要访问的HDFS的URI
19          Configuration conf = new Configuration();
20          // 这里指定使用的是HDFS文件系统
21          conf.set("fs.defaultFS", "hdfs://hadoop:9000");
22          // 通过如下的方式进行客户端身份的设置
23          System.setProperty("HADOOP_USER_NAME", "root");
24          // 通过FileSystem的静态方法获取文件系统客户端对象
25          fs = FileSystem.get(conf);
26      }
27
28      /**
29       * 将本地的爬取的酒店数据和评论原始数据上传到HDFS
```

```
30      * @throws IOException
31      */
32     @Test
33     public void testAddFileToHdfs() throws IOException {
34         // 要上传的文件所在本地路径
35         Path src = new Path("D:/data/hoteldata.csv");
36         // 要上传到HDFS的目标路径 文件名
37         Path dst = new Path("/hdfsdata");
38         // 上传文件方法
39         fs.copyFromLocalFile(src, dst);
40         src = new Path("D:/data/hotelbasic.csv");
41         // 要上传到HDFS的目标路径 文件名
42         dst = new Path("/hdfsdata");
43         // 上传文件方法
44         fs.copyFromLocalFile(src, dst);
45         // 关闭资源
46         fs.close();
47     }
48 }
```

项目中已经包含了数据集文件夹，其中有两个文件：hotelbasic.csv代表酒店基本数据集，hoteldata.csv代表酒店评论数据集。经过以上程序处理，两个文件上传到HDFS中。

上传后使用hadoop命令查看，可以看到多出两个路径，对应评论数据和基本数据，如图11-2所示。

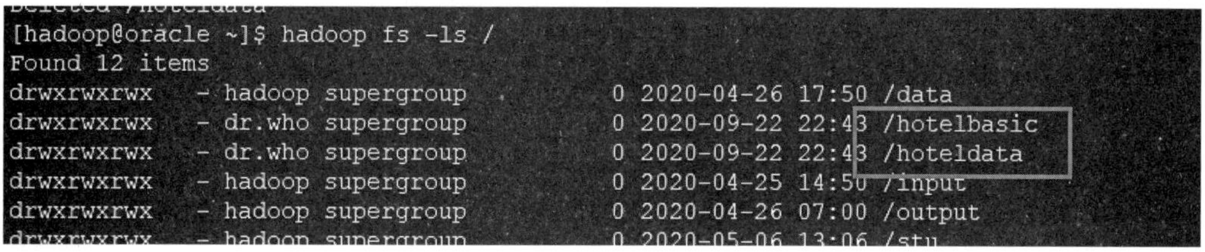

图11-2　Hadoop文件目录

11.3.2　Hadoop 数据清洗

将上传到Hadoop平台的酒店数据进行初步的数据清洗，使其符合大数据分析平台对数据的基本要求。以下是两个数据文件中部分内容：

（1）部分酒店基本数据属性及原始数据如下：

60169364,枫叶酒店式公寓(青岛金沙滩传媒广场店),4.7,黄岛区珠江路588号传媒广场天相公寓4号楼北侧1号101室。（黄岛金沙滩度假区）,hotel_diamond02,经济型

6841087,青岛新天桥快捷宾馆,4.4,市南区肥城路51-1号。（青岛火车站/栈桥/中山路劈柴院）,hotel_diamond02,经济型

4640468,欧圣兰廷公寓(青岛万达东方影都店),4.5,黄岛区滨海大道万达公馆A1区2号楼办理入住。（西海岸度假区）,hotel_diamond02,经济型

4539535,世纪双帆海景度假酒店(青岛城市阳台店),4.4,黄岛区滨海大道1288号那鲁湾1号楼。（西海岸度假区）,hotel_diamond02,经济型

60664769,欧圣兰廷度假公寓(青岛城市阳台店),4.7,黄岛区滨海大道4098号城市阳台风景区世茂悦海13号楼一楼。（西海岸度假区）,hotel_diamond02,经济型

823437,青岛花园大酒店—贵宾楼,4.8,市南区彰化路6号贵宾楼。（五四广场/万象城/奥帆中心/市政府青岛大学）,hotel_diamond04,高档型

```
        43679094,青岛燕岛之星度假公寓,4.7,市南区燕儿岛路15号1楼大厅。（五四广场/万象城/奥帆中心/
市政府）,hotel_diamond02,经济型
        22416789,容锦酒店（青岛台东步行街店）,4.6,市北区延安路129号利群百惠商厦6-7楼。（台东步行
街/啤酒街）,hotel_diamond02,经济型
        17263343,慢居听海酒店（青岛吾悦广场店）,4.8,黄岛区滨海大道2888号梦时代广场17号楼1楼大厅101
室。（黄岛金沙滩度假区）,hotel_diamond02,经济型
        427962,安澜宾舍酒店（青岛东海中路海滨店）,4.4,市南区东海中路30号银海大世界院内。（五四广场/
万象城/奥帆中心/市政府）,hotel_diamond02,经济型
        21122269,青岛蓝朵海景假日公寓,4.8,崂山区秦岭路19号协信中心3号楼28层。（国际会展中心/石老
人海水浴场）,,经济型
```

（2）部分酒店用户评论属性及原始数据如下（评论内容列已去掉）：

```
Elainemimi,枫叶酒店式公寓（青岛金沙滩传媒广场店）,其他,20-Jul,2020/7/6,5
_WeChat268192****,枫叶酒店式公寓（青岛金沙滩传媒广场店）,商务出差,20-Jun,2020/6/28,5
_CFT010000002415****,枫叶酒店式公寓（青岛金沙滩传媒广场店）,朋友出游,20-Jun,2020/6/26,5
_WeChat320342****,枫叶酒店式公寓（青岛金沙滩传媒广场店）,情侣出游,20-Jun,2020/6/20,5
_WeChat320342****,枫叶酒店式公寓（青岛金沙滩传媒广场店）,情侣出游,20-Jun,2020/6/21,5
_WeChat320342****,枫叶酒店式公寓（青岛金沙滩传媒广场店）,情侣出游,20-Jun,2020/6/21,5
M24589****,枫叶酒店式公寓（青岛金沙滩传媒广场店）,独自旅行,20-May,2020/6/4,5
M221656****,青岛新天桥快捷宾馆,家庭亲子,20-Aug,2020/8/27,5
M419190****,青岛新天桥快捷宾馆,家庭亲子,20-Jul,2020/8/10,5
M354977****,青岛新天桥快捷宾馆,独自旅行,20-Jul,2020/7/30,5
M355264****,青岛新天桥快捷宾馆,独自旅行,20-Jul,2020/8/3,5
M311789****,青岛新天桥快捷宾馆,其他,20-Jul,2020/8/3,5
朝生牧者,青岛新天桥快捷宾馆,独自旅行,20-Jul,2020/8/10,5
_WeChat229580****,青岛新天桥快捷宾馆,情侣出游,20-Aug,2020/8/26,4.5
M415985****,青岛新天桥快捷宾馆,独自旅行,20-Jul,2020/8/6,5
_WeChat385461****,青岛新天桥快捷宾馆,朋友出游,20-Aug,2020/8/7,3.8
M416989****,青岛新天桥快捷宾馆,商务出差,20-Jul,2020/8/10,5
M272306****,青岛新天桥快捷宾馆,情侣出游,20-Jun,2020/6/22,5
w风清,青岛新天桥快捷宾馆,情侣出游,20-Jun,2020/8/10,5
_WeChat290262****,青岛新天桥快捷宾馆,朋友出游,20-Jun,2020/8/10,5
杰科1105,青岛新天桥快捷宾馆,家庭亲子,20-Jul,2020/7/22,5
M317651****,青岛新天桥快捷宾馆,家庭亲子,20-Aug,2020/8/18,3.8
通帕蓬廖化,青岛新天桥快捷宾馆,独自旅行,20-Aug,2020/8/19,4.8
M327775****,青岛新天桥快捷宾馆,朋友出游,20-Aug,2020/8/11,5
_WeChat270171****,青岛新天桥快捷宾馆,家庭亲子,20-Aug,2020/8/6,5
平哥儿走天涯,青岛新天桥快捷宾馆,商务出差,20-Jul,2020/8/7,5
_WeChat266127****,青岛新天桥快捷宾馆,朋友出游,20-Aug,2020/8/12,5
```

数据清洗的主要工作：

（1）酒店基本数据集中，酒店星级类型这一列数据叫法不一致，比如有的酒店叫四星级，有的则叫高档型，这里统一做一下处理，将所有的"国家旅游局评定为四星级"替换为"高档型"，将"国家旅游局评定为三星级"替换为"舒适型"，将"国家旅游局评定为二星级"替换为"经济型"，将"国家旅游局评定为五星级"替换为"豪华型"。

（2）由于大数据服务平台这个子模块并不对用户具体评论内容进行情感分析，情感分析是交给情感分析子系统来处理的，所以这里将评论内容数据列去掉。

（3）删除所有空行。

（4）从酒店地址中提取区县名称并替换掉地址那一列内容，为区县酒店分布统计提供标准数据。

具体清洗代码用Java程序编写，其中酒店基本数据清洗及主程序如下：

代码 11.3　HotelBasicClean.java

```java
01 public class HotelBasicClean extends Configured implements Tool {
02     public static void main(String[] args) throws Exception {
03         int run = ToolRunner.run(new HotelBasicClean(), args);
04         System.exit(run);
05     }
06     public int run(String[] args) throws Exception {
07         if (args.length < 2) {
08             System.out.println("参数错误，使用方法：LineCharCountMR <Input> <Output>");
09             ToolRunner.printGenericCommandUsage(System.out);
10             return 1;
11         }
12         Configuration config = getConf();
13         FileSystem fs = FileSystem.get(config);
14         Path dest = new Path(args[1]);
15         if (fs.exists(dest)) {
16             fs.delete(dest, true);
17         }
18         Job job = Job.getInstance(config, "LineChar");
19         job.setJarByClass(getClass());
20         job.setMapperClass(LineMapper.class);
21         job.setOutputKeyClass(Text.class);
22         job.setOutputValueClass(NullWritable.class);
23         FileInputFormat.addInputPath(job, new Path(args[0]));
24         FileOutputFormat.setOutputPath(job, dest);
25         boolean b = job.waitForCompletion(true);
26         return b ? 0 : 1;
27     }
28     //注意最后一个参数为NullWritable,可以理解为Null
29     public static class LineMapper extends Mapper<LongWritable, Text, Text, NullWritable> {
30         Text ntxt=new Text();
31         @Override
32         protected void map(LongWritable key, Text value, Context context) throws IOException, InterruptedException {
33             String line = value.toString();
34             if (StringUtils.isBlank(line)) {//去掉空行
35                 return;
36             }
37             String[] arr=line.split(",");
38             String a=arr[-3];//地址
39             //提取县、市、区替换地址
40             if(a.contains("平度")) arr[-3]= "平度市";
41             else if(a.contains("市南")) arr[-3]= "市南区";
42             else if(a.contains("市北")) arr[-3]= "市北区";
43             else if(a.contains("李沧")) arr[-3]= "李沧区";
44             else if(a.contains("城阳")) arr[-3]= "城阳区";
45             else if(a.contains("崂山")) arr[-3]= "崂山区";
46             else if(a.contains("黄岛"))arr[-3]= "黄岛区";
47             else if(a.contains("即墨")) arr[-3]= "即墨市";
48             else if(a.contains("胶州")) arr[-3]= "胶州市";
49             else if(a.contains("莱西"))arr[-3]= "莱西市";
50             else arr[-3] ="其他";
51             //获取最后一列酒店详情
```

```
52          String b=arr[-1];
53          if(b.equals("国家旅游局评定为四星级")) arr[-1]="高档型";
54          else if(b.equals("国家旅游局评定为五星级")) arr[-1]="豪华型";
55          else if(b.equals("国家旅游局评定为三星级")) arr[-1]="舒适型";
56          else if(b.equals("国家旅游局评定为二星级")) arr[-1]="经济型";
57          else;
58          //新数组重新组成字符串，用逗号隔开
59          String nline="";
60          for (String ele:arr) {
61              nline+=ele+",";
62          }
63          nline=nline.substring(0,nline.length()-1);
64          ntxt.set(nline);
65
66          context.write(ntxt, NullWritable.get());
67        }
68      }
69
70 }
```

酒店评论数据清洗的代码如下：

代码 11.4　HotelDataClean.java

```
01 public class HotelDataClean extends Configured implements Tool {
02     public static void main(String[] args) throws Exception {
03         int run = ToolRunner.run(new HotelDataClean(), args);
04         System.exit(run);
05     }
06     public int run(String[] args) throws Exception {
07         if (args.length < 2) {
08             System.out.println("参数错误，使用方法：LineCharCountMR <Input> <Output>");
09             ToolRunner.printGenericCommandUsage(System.out);
10             return 1;
11         }
12         Configuration config = getConf();
13         FileSystem fs = FileSystem.get(config);
14         Path dest = new Path(args[1]);
15         if (fs.exists(dest)) {
16             fs.delete(dest, true);
17         }
18         Job job = Job.getInstance(config, "LineChar");
19         job.setJarByClass(getClass());
20         job.setMapperClass(LineMapper.class);
21         job.setOutputKeyClass(Text.class);
22         job.setOutputValueClass(NullWritable.class);
23         FileInputFormat.addInputPath(job, new Path(args[0]));
24         FileOutputFormat.setOutputPath(job, dest);
25         boolean b = job.waitForCompletion(true);
26         return b ? 0 : 1;
27     }
28     //注意最后一个参数为NullWritable,可以理解为Null
29     public static class LineMapper extends Mapper<LongWritable, Text, Text, NullWritable> {
30         Text ntxt=new Text();
```

```
31         @Override
32         protected void map(LongWritable key, Text value, Context context) throws
   IOException, InterruptedException {
33             String line = value.toString();
34             if (StringUtils.isBlank(line)) {//去掉空行
35                 return;
36             }
37             String[] arr=line.split(",");
38
39             //去掉最后一列
40             String nline="";
41             for (int i=0;i<arr.length-1;i++) {
42                 nline+=arr[i]+",";
43             }
44             nline=nline.substring(0,nline.length()-1);
45             ntxt.set(nline);
46
47             context.write(ntxt, NullWritable.get());
48         }
49     }
50
51 }
```

清洗后的数据分别存放到HDFS的两个不同目录下，hotelbasic存放基本数据，hoteldata存放评论数据，如图11-3所示。

图11-3　清洗后的Hadoop文件系统目录

11.3.3　构建 Hive 数据仓库表

酒店大数据分析服务平台创建了两个外部表，分别对应酒店基本信息表和酒店用户评论表。

Hive内部表是分别对关于酒店数据的5个方面指标进行分析得到的结果，最终内部表数据会导入到MySQL中。各表结构设计如表11-1~表11-5所示。

表 11-1　出游类型统计表设计

字段编号	字段名称	数据类型	约束	字段描述
1	triptype	varchar	PK	出游类型
2	num	int		数量

表 11-2　用户满意度分布统计表设计

字段编号	字段名称	数据类型	约束	不是 Nulls	字段描述
1	bad	int		☑	不满意
2	good	int		☑	满意
3	excellent	int		☑	非常满意

表 11-3 酒店星级分布统计表

字段编号	字段名称	数据类型	约束	不是 Nulls	字段描述
1	stardetail	varchar	PK	☑	酒店星级
2	nums	int		☑	数量

表 11-4 酒店评论数量统计表

字段编号	字段名称	数据类型	约束	字段描述
1	hotel_name	varchar	PK	酒店名称
2	num	int		数量

表 11-5 各区县酒店数量和用户推荐评分表

字段编号	字段名称	数据类型	约束	字段描述
1	area_name	varchar	PK	区县名称
2	num	int		数量
3	recommend	int		推荐平均评分

1. 基于上传的酒店用户评论数据hoteldata.csv创建Hive外部表

```
create external table hotel_data(user_name string,hotel_name string,trip_type string,time1 string,time2 string ,user_score double)
ROW FORMAT DELIMITED FIELDS TERMINATED BY ',' LOCATION '/hoteldata';
```

这样数据就会自动加载/hoteldata下面的数据。

2. 基于上传的酒店基本数据hotelbasic.csv创建Hive表

```
create external table hotel_basic(id string,name string,score double,commentnum int,recommend int,address string,star  string,stardetail string)
ROW FORMAT DELIMITED FIELDS TERMINATED BY ',' LOCATION '/hotelbasic';
```

这样数据就会自动加载/hotelbasic下面的数据。

3. 基于Hive外部表统计以下数据形成内部表,并将分析结果导出到MySQL数据库

(1) 用户印象统计

用户住过酒店发表评论同时也可以打分,以下是根据用户对该地区或城市的酒店总体评分情况来统计用户总体印象。评分4.5~5分为优良,3.5~4.5为良好,3.5以下为差,统计酒店用户评分等级比例。提前建好Hive内部表score_stat表结构,并覆盖式插入数据到该表。

```
create table score_stat(bad int,good int,excellent int)
```

插入数据:

```
insert overwrite table score_stat select  count(case when user_score<3.5 then 1 else null end) as 'bad',count(case when user_score>=3.5 and user_score<4.5 then 1 else null end) as 'good',
    count(case when user_score>=4.5 then 1 else null end) as 'excellent'  from hotel_data
```

将数据导出到MySQL中:

```
sqoop export --connect jdbc:mysql://192.168.1.110:3306/test --username root --password root --table score_stat --fields-terminated-by '\001' --export-dir '/user/hive/warehouse/score_stat '
```

（2）统计在线评论数最多的十大酒店、十大网络人气酒店

一般情况下一家酒店的评论数量能代表这家酒店的人气，这里统计的是酒店名称和评论数目前十的酒店，设计评论数目表结构：

```
create table comments_stat(hotel_name string,nums int)
insert overwritetable comments_stat select hotel_name,count(1) as nums from hotel_data group by hotel_name
```

将该数据导入到MySQL。Sqoop导入命令：

```
sqoop export --connect jdbc:mysql://192.168.1.110:3306/test --username root --password root --table comments_stat --fields-terminated-by '\001' --export-dir '/user/hive/warehouse/ comments_stat '
```

（3）不同旅游类型占比统计

```
create table triptype_stat(triptype string,nums bigint)
insert overwrite table triptype_stat select trip_type,count(1) as nums from hotel_data group by trip_type
```

设计MySQL表结构：

```
create table triptype_stat(triptype varchar(33),nums int)
```

将该数据导入到MySQL。Sqoop导入命令：

```
sqoop export --connect jdbc:mysql://192.168.1.110:3306/test --username root --password root --table triptype_stat --fields-terminated-by '\001' --export-dir '/user/hive/warehouse/triptype_stat '
```

（4）酒店星级分布情况统计

设计酒店星级和数量结构，如下：

```
create table star_stat(stardetail string,nums int)
insert overwrite table star_stat select stardetail,count(1) from hotel_basic group by stardetail
```

设计MySQL表结构：

```
create table star_stat(stardetail varchar(33),nums int)
```

将该数据导入到MySQL。Sqoop导入命令：

```
sqoop export --connect jdbc:mysql://192.168.1.110:3306/test --username root --password root --table star_stat --fields-terminated-by '\001' --export-dir '/user/hive/warehouse/star_stat '
```

（5）城市不同区的酒店数量分布情况，以热力图方式呈现

```
create table area_stat(area_name string,nums int,recommend int)
```

插入数据：

```
insert overwrite table area_stat  select address,round(avg(recommend),2)  from
hotel_basic group by regexp_extract(address,'(.+?区|市|县)(.*)',1);
```

以上查询语句是使用正则表达式从酒店数据中的酒店地址中提取区、市、县的名称。

设计MySQL表结构：

```
create table area_stat (area_name varchar(33),nums int,recommend int)
```

将该数据导入到MySQL。Sqoop导入命令：

```
sqoop export --connect jdbc:mysql://192.168.1.110:3306/test --username root
--password root --table area_stat --fields-terminated-by '\001' --export-dir
'/user/hive/warehouse/area_stat '
```

将以上命令形成Shell脚本，命名为bigdata.sh，通过运行脚本来构建数据仓库并取得分析结果。完整的脚本代码如下：

代码 11.5　bigdata.sh

```
01  #!/bin/sh
02
03  #创建外部表
04  echo "基于上传的hoteldata.csv创建Hive外部表"
05  hive -e "create external table if not exists  hotel_data(user_name string,hotel_name
       string,trip_type string,time1 string,time2 string ,user_score double)
06  ROW FORMAT DELIMITED FIELDS TERMINATED BY ',' LOCATION '/hoteldata'"
07  echo "基于上传的hotelbasic.csv创建Hive表"
08  hive -e "create external table if not exists  hotel_basic(id string,name string,score
       double,commentnum int,recommend int,address string,star  string,stardetail string)
09  ROW FORMAT DELIMITED FIELDS TERMINATED BY ',' LOCATION '/hotelbasic'"
10
11  #创建Hive内部表
12  echo "================================创建Hive内部表
       ================================"
13  #1.来该地区用户的旅游类型统计
14  echo "1.用户的旅游类型统计表........................................."
15  hive -e "create table triptype_stat(triptype string,nums bigint)"
16  hive -e "insert overwrite table triptype_stat  select trip_type,count(1) as nums from
       hotel_data group by trip_type"
17  #将数据导入到MySQL
18  echo "导入数据到MySQL开始"
19  sqoop export --connect
    jdbc:mysql://192.168.1.110:3306/hotel?useUnicode=true\&characterEncoding=utf-8
    --username root --password 123456 --table triptype_stat --fields-terminated-by
    '\001' --export-dir '/user/hive/warehouse/triptype_stat'
20  echo "导入数据到MySQL结束"
21  hive -e "drop table  triptype_stat;"
22  echo "导入MySQL之后删除Hive内部表triptype_stat........................................."
23
24  #2.根据用户对该地区或城市的酒店总体评分情况来统计用户总体印象，用户评分统计
25  echo "2. 用户对该地区或城市的酒店总体评分情况来统计用户总体印象........................."
26  hive -e "create table score_stat(bad int,good int,excellent int)"
27  hive -e "insert overwrite table score_stat select  count(case when user_score<3.5 then
       1 else null end)  as bad,count(case when user_score>=3.5 and user_score<4.5 then 1
       else null end)  as good,count(case when user_score>=4.5 then 1 else null end)  as
       excellent  from hotel_data"
```

```bash
28  #将数据导入到MySQL
29  echo "导入数据到MySQL开始"
30  sqoop export --connect
    jdbc:mysql://192.168.1.110:3306/hotel?useUnicode=true\&characterEncoding=utf-8
    --username root --password 123456 --table score_stat --fields-terminated-by '\001'
    --export-dir '/user/hive/warehouse/score_stat'
31  echo "导入数据到MySQL结束"
32  hive -e "drop table  score_stat;"
33  echo "导入MySQL之后删除Hive内部表score_stat..................................."
34
35  #3.十大网络人气酒店,酒店名称和用户评论数量
36  echo "3.十大网络人气酒店........................................."
37  hive -e "create table comments_stat(hotel_name string,nums int)"
38  hive -e "insert overwrite table comments_stat select hotel_name,count(1) as nums from
     hotel_data group by hotel_name"
39  #将数据导入到MySQL
40  echo "导入数据到MySQL开始"
41  sqoop export --connect
    jdbc:mysql://192.168.1.110:3306/hotel?useUnicode=true\&characterEncoding=utf-8
    --username root --password 123456 --table comments_stat --fields-terminated-by
    '\001' --export-dir '/user/hive/warehouse/comments_stat'
42  echo "导入数据到MySQL结束"
43  hive -e "drop table  comments_stat;"
44  echo "导入MySQL之后删除Hive内部表comments_stat..................................."
45
46  #4.酒店星级分布情况统计,设计酒店星级和数量结构
47  echo "4.酒店星级分布情况统计........................................."
48  hive -e "create table star_stat(stardetail string,nums int)"
49  hive -e "insert overwrite table star_stat  select stardetail,count(1) from hotel_basic
     group by stardetail"
50  #将数据导入到MySQL
51  echo "导入数据到MySQL开始"
52  sqoop export --connect
    jdbc:mysql://192.168.1.110:3306/hotel?useUnicode=true\&characterEncoding=utf-8
    --username root --password 123456 --table star_stat  --fields-terminated-by '\001'
    --export-dir '/user/hive/warehouse/star_stat'
53  echo "导入数据到MySQL结束"
54  hive -e "drop table  star_stat;"
55  echo "导入MySQL之后删除Hive内部表star_stat..................................."
56  #5.各地区酒店数量统计
57  echo "各地区酒店数量统计........................................."
58  hive -e "create table area_stat(area_name string,nums int,recommend int)"
59  hive -e "insert overwrite table area_stat  select
     address ,count(1),round(avg(recommend),2)  from hotel_basic group by  address"
60  #将数据导入到MySQL
61  echo "导入数据到MySQL开始"
62  sqoop export --connect
    jdbc:mysql://192.168.1.110:3306/hotel?useUnicode=true\&characterEncoding=utf-8
    --username root --password 123456 --table area_stat  --fields-terminated-by '\001'
    --export-dir '/user/hive/warehouse/area_stat'
63  echo "导入数据到MySQL结束"
64  hive -e "drop table  area_stat;"
65  echo "导入MySQL之后删除Hive内部表area_stat..................................."
```

运行过程如图11-4所示。

```
[hadoop@oracle ~]$ ./bigdata.sh
基于上传的hoteldata.csv创建hive外部表

Logging initialized using configuration in jar:file:/home/hadoop/hive/lib/hive-c
ommon-1.2.2.jar!/hive-log4j.properties
OK
Time taken: 0.985 seconds
基于上传的hotelbasic.csv创建hive表

Logging initialized using configuration in jar:file:/home/hadoop/hive/lib/hive-common-1.2.2.jar!/hive-lo
g4j.properties
OK
Time taken: 0.834 seconds
=============================创建hive内部表==============================
1.用户的旅游类型统计表.............................................

Logging initialized using configuration in jar:file:/home/hadoop/hive/lib/hive-common-1.2.2.jar!/hive-lo
g4j.properties
OK
Time taken: 1.165 seconds

Logging initialized using configuration in jar:file:/home/hadoop/hive/lib/hive-common-1.2.2.jar!/hive-lo
g4j.properties
Query ID = hadoop_20200922125745_de2fdb3d-3757-469f-922b-44448e866c1f
Total jobs = 1
Launching Job 1 out of 1
Number of reduce tasks not specified. Estimated from input data size: 1
In order to change the average load for a reducer (in bytes):
  set hive.exec.reducers.bytes.per.reducer=<number>
In order to limit the maximum number of reducers:
  set hive.exec.reducers.max=<number>
In order to set a constant number of reducers:
  set mapreduce.job.reduces=<number>
Starting Job = job_1600749800332_0001, Tracking URL = http://oracle:8088/proxy/application_1600749800332
_0001/
Kill Command = /home/hadoop/hadoop-2.7.7/bin/hadoop job  -kill job_1600749800332_0001
Hadoop job information for Stage-1: number of mappers: 1; number of reducers: 1
```

图11-4　构建Hive数据表脚本执行过程

通过以上脚本的运行，根据Hive仓库外部表数据，进行城市酒店数据评分、用户印象、各区、县酒店数量和评分、网络人气、酒店星级、来此地游客的旅游目的类型统计等数据分析，每一项统计都构建Hive内部表。

MySQL数据库对应的表结构需要提前创建好，以方便Hive将内部表数据导出到MySQL，建表脚本如下：

```
create table triptype_stat(triptype varchar(33),nums int);
create table score_stat(bad int,good int,excellent int);
create table comments_stat(hotel_name varchar(200),nums int);
create table star_stat(stardetail varchar(200),nums int)
create table area_stat(area_name varchar(200),nums int,recommend int)
```

11.3.4 Sqoop 数据导入与导出

Hive表数据导出到MySQL需要用到Sqoop组件，这里补充说明一下。

Sqoop是一个数据迁移工具。Sqoop使用非常简单，其整合了Hive、HBase和Oozie，通过 MapReduce任务来传输数据，从而提供并发特性和容错。Sqoop由于是将数据导入到HDFS中，所以需要依赖于Hadoop，即前提是Hadoop已经安装且正确配置。

Sqoop主要通过JDBC和关系数据库进行交互。理论上支持JDBC的数据库都可以使用Sqoop与HDFS进行数据交互，比如将数据库中的数据导入到HDFS，或者将HDFS中的数据导入到数据库中。

1. import导入数据到HDFS

import命令用于将数据库中的数据导入到HDFS。其中--table参数用于将一个表中的数据全部导入到HDFS中去。

（1）--table指定导出的表

```bash
#!/bin/bash
./sqoop import \
--connect \
jdbc:mysql://192.168.1.110:3306/opt?characterEncoding=UTF-8 \
--username root \
--password 1234 \
--table studs \           #指定表名
-m 2 \                    #指定mapper的个数，不能超过集群节点的数量，默认为4
--split-by "id" \         #只要-m不是1必须要指定分组的字段名称
--where "age>100 and sex='1'" \   #指定where条件，可以使用""""双引号
--target-dir /out001      #指定导入到HDFS以后目录
```

默认导出到HDFS的数据以英文逗号","分开，如下所示：

```
$ hdfs dfs -cat /out001/*
2a56b3536b544f289ba79b2b5c1196c4,Jerry,89e4a3e..3ee3e03946d85d
cc645dc7811740fc9856b1c7c8e19e89,Alex,c924e3..5a0487e23207986189d
U001,Jack,1234
U002,Mike,1234
```

可以使用--fields-terminated-by参数指定分隔符号，如--fields-terminated-by "\t"将分隔符号设置为制表符。

（2）--query指定查询语句

如果在import中已经使用了--query语句，则--where和--table将被忽略。在--query所指定的语句中，必须将$CONDITIONS作为条件添加到where子句中。如果--query使用后面""""双引号，则应该使用\$CONDITIONS。注意前面的\（斜线）。

```bash
#!/bin/bash
sqoop import \
--connect jdbc:mysql://192.168.1.110:3306/qlu?characterEncoding=UTF-8 \
--username root \
--password 1234 \
#注意以下使用的SQL语句，如果使用""""双引号，则必须要添加\在$CONDITIONS前面
--query "select name,sex,age,addr from studs where sex='0' and addr like '山东%' and \$CONDITIONS" \
--split-by "name" \
--fields-terminated-by "\t" \  #使用制表符号进行数据分隔
--target-dir /out002 \
-m 2
```

--query参数的SQL可以写得很复杂，如下面的示例，将是一个关联的查询语句：

```bash
#!/bin/bash
sqoop import \
--connect jdbc:mysql://192.168.1.110:3306/studs?characterEncoding=UTF-8 \
--username root \
--password 1234 \
--query \
"select s.stud_id,s.stud_name as sname,c.course_name as cname \
from studs s inner join sc on s.stud_id=sc.sid \
    inner join courses c on c.course_id=sc.cid where \$CONDITIONS" \
--target-dir /out004 \
```

```
--split-by "s.stud_id" \    #根据某个列进行分组
-m 2
```

2. export导出数据到数据库

使用sqoop export命令可以将HDFS数据导出到关系数据库MySQL中去。

```
#!/bin/bash
sqoop export \
- -connect \
 jdbc:mysql://192.168.1.110:3306/opt?characterEncoding=UTF-8 \
--username root \
--password 1234 \
--export-dir /out001 \      #指定导出的目录
--table "studs" \           #指定HDFS中数据与数据库中表列的对应关系
--columns "stud_id,stud_age,stud_name" \#指定HDFS中数据进行分隔
- -fields-terminated-by "\t" \
-m 2
```

11.3.5 数据可视化开发

本小节将11.3.3节中的分析结果，即MySQL中的数据，以图表的方式呈现到Web网页上。

这里采用Spring Boot+MyBatis+MySQL构建可视化项目，本小节具体代码可以参见配套资源中本章项目文件夹下的HotelVisualization。

图表展示使用ECharts，ECharts是一个纯JavaScript图表库，底层依赖于轻量级的Canvas类库ZRender，基于BSD开源协议，是一款非常优秀的可视化前端框架。它提供直观、生动、可交互、可高度个性化定制的数据可视化图表。其创新的拖曳重计算、数据视图、值域漫游等特性大大增强了用户体验，赋予了用户对数据进行挖掘、整合的能力。它支持折线图（区域图）、柱状图（条状图）、散点图（气泡图）、K线图、饼图（环形图）等。

首先，新建一个Maven项目，引入需要使用的框架，如Spring Boot、MyBatis等，具体pom.xml配置信息如下：

代码 11.6　pom.xml

```
01  <?xml version="1.0" encoding="UTF-8"?>
02  <project xmlns="http://maven.apache.org/POM/4.0.0"
      xmlns:xsi="http://www.w3.org/2001/XMLSchema-instance"
03       xsi:schemaLocation="http://maven.apache.org/POM/4.0.0
    http://maven.apache.org/xsd/maven-4.0.0.xsd">
04     <modelVersion>4.0.0</modelVersion>
05     <parent>
06         <groupId>org.springframework.boot</groupId>
07         <artifactId>spring-boot-starter-parent</artifactId>
08         <version>2.1.6.RELEASE</version>
09         <relativePath/> <!-- lookup parent from repository -->
10     </parent>
11     <groupId>com.zjp.echartsdemo</groupId>
12     <artifactId>echartsdemo</artifactId>
13     <version>0.0.1-SNAPSHOT</version>
14     <name>echartsdemo</name>
15     <description>Demo project for Spring Boot</description>
16
```

```xml
17  <properties>
18      <java.version>1.8</java.version>
19  </properties>
20
21
22  <dependencies>
23      <dependency>
24          <groupId>org.springframework.boot</groupId>
25          <artifactId>spring-boot-starter-web</artifactId>
26      </dependency>
27
28      <dependency>
29          <groupId>org.springframework.boot</groupId>
30          <artifactId>spring-boot-starter-test</artifactId>
31          <scope>test</scope>
32      </dependency>
33      <dependency>
34          <groupId>org.webjars.bower</groupId>
35          <artifactId>echarts</artifactId>
36          <version>4.2.1</version>
37      </dependency>
38      <dependency>
39          <groupId>org.webjars</groupId>
40          <artifactId>jquery</artifactId>
41          <version>3.4.1</version>
42      </dependency>
43      <dependency>
44          <groupId>org.mybatis.spring.boot</groupId>
45          <artifactId>mybatis-spring-boot-starter</artifactId>
46          <version>2.0.1</version>
47      </dependency>
48      <dependency>
49          <groupId>mysql</groupId>
50          <artifactId>mysql-connector-java</artifactId>
51          <scope>runtime</scope>
52      </dependency>
53      <dependency>
54          <groupId>com.github.pagehelper</groupId>
55          <artifactId>pagehelper-spring-boot-starter</artifactId>
56          <version>1.2.5</version>
57      </dependency>
58      <!-- alibaba的druid数据库连接池 -->
59      <dependency>
60          <groupId>com.alibaba</groupId>
61          <artifactId>druid-spring-boot-starter</artifactId>
62          <version>1.1.9</version>
63      </dependency>
64      <!-- 引入Thymeleaf依赖 -->
65      <dependency>
66          <groupId>org.springframework.boot</groupId>
67          <artifactId>spring-boot-starter-thymeleaf</artifactId>
68      </dependency>
69      <dependency>
70          <groupId>org.springframework.boot</groupId>
71          <artifactId>spring-boot-devtools</artifactId>
72          <optional>true</optional>
```

```
73            </dependency>
74
75        </dependencies>
76
77        <build>
78
79            <plugins>
80                <plugin>
81                    <groupId>org.springframework.boot</groupId>
82                    <artifactId>spring-boot-maven-plugin</artifactId>
83                </plugin>
84            </plugins>
85        </build>
86
87    </project>
```

然后，配置Spring Boot的配置文件，定义数据库连接池配置以及Web服务器端口配置，具体配置信息如下：

代码 11.7　application.properties

```
01  server.port=8088
02  #数据库连接池配置
03  spring.datasource.name=zjptest
04  spring.datasource.type=com.alibaba.druid.pool.DruidDataSource
05  spring.datasource.druid.filters=stat
06  spring.datasource.druid.driver-class-name=com.mysql.jdbc.Driver
07  spring.datasource.druid.url=jdbc:mysql://192.168.1.110:3306/hotel?useUnicode=true
       &characterEncoding=UTF-8&allowMultiQueries=true&serverTimezone=UTC
08  spring.datasource.druid.username=root
09  spring.datasource.druid.password=123456
10  pring.datasource.druid.initial-size=1
11  spring.datasource.druid.min-idle=1
12  spring.datasource.druid.max-active=20
13  spring.datasource.druid.max-wait=6000
14  spring.Hadatasource.druid.time-between-eviction-runs-millis=60000
15  spring.datasource.druid.min-evictable-idle-time-millis=300000
16  spring.datasource.druid.validation-query=SELECT 'x'
17  spring.datasource.druid.test-while-idle=true
18  spring.datasource.druid.test-on-borrow=false
19  spring.datasource.druid.test-on-return=false
20  spring.datasource.druid.pool-prepared-statements=false
21  spring.datasource.druid.max-pool-prepared-statement-per-connection-size=20
22  #mybatis配置
23  mybatis.mapper-locations=classpath:mapper/*.xml
24  mybatis.type-aliases-package= com.dpzhou.echartsdemo.echartsdemo.entity
25  #thymeleaf配置
26  spring.thymeleaf.cache=false
27  spring.thymeleaf.prefix=classpath:/templates/
28  spring.thymeleaf.suffix=.html
29  spring.thymeleaf.mode=HTML5
30  spring.thymeleaf.encoding=UTF-8
31  spring.thymeleaf.check-template-location=true
```

根据项目需求，需要开发5个可视化统计图，因为项目使用了MyBatis框架，所以可以将SQL直接配置在Mapper文件中，详细配置信息如下：

代码 11.8　CommonMapper.xml

```xml
01 <?xml version="1.0" encoding="UTF-8"?>
02 <!DOCTYPE mapper PUBLIC "-//mybatis.org//DTD Mapper 3.0//EN"
     "http://mybatis.org/dtd/mybatis-3-mapper.dtd">
03 <mapper namespace="com.zjp.echartsdemo.echartsdemo.dao.CommonMapper">
04
05
06    <select id="selectTripType" parameterType="java.lang.String" resultType="map">
07        select
08         triptype,nums
09        from triptype_stat order by nums desc limit 0,7
10    </select>
11    <select id="selectCommentsStat" parameterType="java.lang.String"
   resultType="map">
12        select
13         hotel_name,nums
14        from comments_stat order by nums desc limit 0,10
15    </select>
16    <select id="selectScoreStat" parameterType="java.lang.String" resultType="map">
17        select
18         *
19        from score_stat
20    </select>
21    <select id="selectStarStat" parameterType="java.lang.String" resultType="map">
22        select
23         stardetail,nums
24        from star_stat order by nums desc limit 0,4
25    </select>
26    <select id="selectAreaStat" parameterType="java.lang.String" resultType="map">
27        select * from area_stat where area_name in('市南区','市北区','黄岛区','即墨区','
   城阳区','崂山区','李沧区') order by nums desc limit 0,10
28    </select>
29 </mapper>
```

前端异步请求数据采用Ajax技术，后台查询数据以json格式返回，通过ECharts在页面加载返回json数据显示结果。

定义view.html页面，划分div分别显示对应的统计图表，页面JS脚本采用Ajax实现无刷新交互，具体页面代码如下：

代码 11.9　view.html

```html
01 <!-- 为 ECharts 准备一个具备大小（宽高）的 DOM -->
02 <h1 class="aa">青岛市酒店大数据可视化</h1>
03 <div id="main" style="width:
   500px;height:400px;position:absolute;top:100px"></div><!--用户的旅游类型统计-->
04 <div id="main2" style="width:
   500px;height:400px;position:absolute;top:100px;left:550px"></div><!--地区酒店数量
   统计-->
```

```
05  <div id="main3" style="width: 500px;height:400px;position:absolute;top:100px;
    left:1050px"></div><!--用户印象评分等级比例统计-->
06  <a href="/echarts" >Goto基于用户评论统计分析可视化</a>
07  <script type="text/javascript">
08      // 基于准备好的dom，初始化ECharts实例
09      var myChart = echarts.init(document.getElementById('main'));
10      // 新建productName与nums数组来接收数据
11      var triptypes = [];
12      var nums = [];
13      var json = {};
14      var datatemp = [];
15
16      //旅游类型统计
17      $.ajax({
18          type:"GET",
19          url:"/triptypestat",
20          dataType:"json",
21          async:false,
22          success:function (result) {
23              json = result;
24              for (var i = 0; i < result.length; i++){
25                  triptypes.push(result[i].triptype);
26                  nums.push(result[i].nums);
27                  var ob = {name:"",value:""};
28                  ob.name = result[i].triptype;
29                  ob.value = result[i].nums;
30                  datatemp.push(ob);
31              }
32
33          },
34          error :function(errorMsg) {
35              alert("获取后台数据失败！");
36          }
37      });
38
39      // 指定图表的配置项和数据
40      var option = {
41          title: {
42              text: '用户的旅游类型统计'
43          },
44          tooltip: {},
45          legend: {
46              data:['人次']
47          },
48          xAxis: {
49              //结合
50              axisLabel: {
51                  interval:0,
52                  rotate:20
53              },
54              data: triptypes
55          },
56
57          yAxis: {},
58          series: [{
```

```
59              name: '旅游类型',
60              type: 'bar',
61              //结合
62              data: nums
63          }]
64      };
65
66      // 使用刚指定的配置项和数据显示图表
67      myChart.setOption(option);
68
69
70      //加载地图数据
71      //旅游类型统计
72      $.ajax({
73          type:"GET",
74          url:"/areastat",
75          dataType:"json",
76          async:false,
77          success:function (result) {
78              json = result;
79              for (var i = 0; i < result.length; i++){
80                  var ob = {name:"",value:""};
81                  ob.name = result[i].area_name;
82                  ob.value = result[i].nums;
83                  datatemp.push(ob);
84              }
85
86          },
87          error :function(errorMsg) {
88              alert("获取后台数据失败!");
89          }
90      });
91      var myChart2 = echarts.init(document.getElementById('main2'));
92
93      option = {
94          title: {
95              text: '各地区酒店数量统计'
96          },
97          tooltip: {
98              formatter:function(params,ticket, callback){
99                  return params.seriesName+'<br />'+params.name+': '+params.value
100             }
101         },
102
103         visualMap: {
104             min: 0,
105             max: 1500,
106             left: 'left',
107             top: 'bottom',
108             text: ['高','低'],
109             inRange: {
110                 color: ['#e5e0e0', '#490104']
111             },
112             show:true
113         },
```

```
            geo: {
                map: 'QD',
                roam: false,
                zoom:1.23,
                label: {
                    normal: {
                        show: true,
                        fontSize:'10',
                        color: 'rgba(0,0,0,0.7)'
                    }
                },
                itemStyle: {
                    normal:{
                        borderColor: 'rgba(0, 0, 0, 0.2)'
                    },
                    emphasis:{
                        areaColor: '#F3B329',
                        shadowOffsetX: 0,
                        shadowOffsetY: 0,
                        shadowBlur: 20,
                        borderWidth: 0,
                        shadowColor: 'rgba(0, 0, 0, 0.5)'
                    }
                },

            series : [
                {
                    name: '酒店数量',
                    type: 'map',
                    geoIndex: 0,
                    data:datatemp
                }

            ]
        };

$.getJSON('js/370200.json', function (geoJson) {
    myChart2.hideLoading();
    echarts.registerMap('QD', geoJson);
    myChart2.setOption(option);
})


//酒店星级分布情况统计
var myChart3 = echarts.init(document.getElementById('main3'));

var option3 = {
    title : {
        text: '酒店星级分布情况统计',
        subtext: '',
        x:'center'
    },
    tooltip : {
        trigger: 'item',
        formatter: "{a} <br/>{b} : {c} ({d}%)"
```

```
170            },
171        legend: {
172            orient: 'vertical',
173            left: 'left',
174        },
175        series : [
176            {
177                name: '酒店类型',
178                type: 'pie',
179                radius : '55%',
180                center: ['50%', '60%'],
181                data:(function () {
182
183                    var datas = [];
184                    $.ajax({
185                        type:"POST",
186                        url:"/starstat",
187                        dataType:"json",
188                        async:false,
189                        success:function (result) {
190
191                            for (var i = 0; i < result.length; i++){
192                                datas.push({
193                                    "value":result[i].nums, "name":result[i].stardetail
194                                })
195                            }
196
197                        }
198                    })
199                    return datas;
200
201                })(),
202                itemStyle: {
203                    emphasis: {
204                        shadowBlur: 10,
205                        shadowOffsetX: 0,
206                        shadowColor: 'rgba(0, 0, 0, 0.5)'
207                    }
208                }
209            }
210        ]
211    };
212    myChart3.setOption(option3);
213</script>
```

具体Spring Boot中Controller、Service和Dao层的代码不在本书讨论范围之内，可以参见配套的项目源码。其中基于酒店基本数据的统计分析部分包括：用户旅游类型分析、各地区酒店数量统计、酒店星级情况统计3个，可视化结果如图11-5所示。

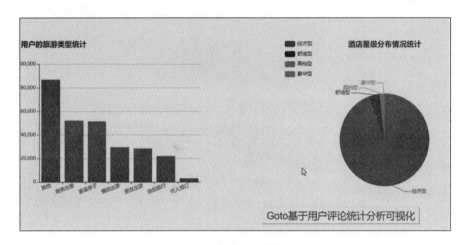

图11-5 酒店基本数据统计图

基于酒店用户评论数据的统计分析部分包括：网络人气酒店和用户满意度统计，可视化结果如图11-6所示。

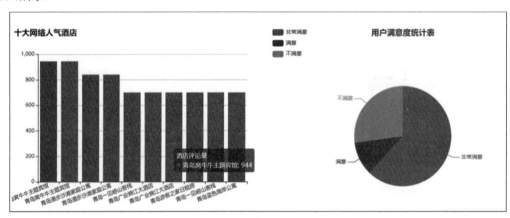

图11-6 用户评论数据统计图